Textbook of Plant Diseases

Textbook of Plant Diseases

– Author –
Bibhuti Dutta Jha

BIO - GREEN
BOOKS

BIO-GREEN BOOKS
New Delhi - 110 002

ISBN: 978-93-90438-10-5 (Hardback)

Published by : **BIO-GREEN BOOKS**
4736/23, Ansari Road, Darya Ganj,
New Delhi – 110 002
Phone: 011-23245578, 23244987
E-mail: biogreenbooks@gmail.com

Printed at **:** **Replika Press Pvt. Ltd.**

Preface

Plant pathology is the scientific study of plant diseases caused by pathogens (infectious organisms) and environmental conditions. Organisms that cause infectious disease include fungi, oomycetes, bacteria, viruses, viroids, virus-like organisms, phytoplasmas, protozoa, nematodes and parasitic plants. Not included are ectoparasites like insects, mites, vertebrate, or other pests that affect plant health by consumption of plant tissues. Plant pathology also involves the study of pathogen identification, disease etiology, disease cycles, economic impact, plant disease epidemiology, plant disease resistance, how plant diseases affect humans and animals, pathosystem genetics, and management of plant diseases.

The majority of phytopathogenic fungi belong to the Ascomycetes and the Basidiomycetes. The fungi reproduce both sexually and asexually via the production of spores and other structures. Spores may be spread long distances by air or water, or they may be soil borne. Many soil inhabiting fungi are capable of living saprotrophically, carrying out the part of their lifecycle in the soil. These are known as facultative saprotrophs. Fungal diseases may be controlled through the use of fungicides and other agriculture practices, however new races of fungi often evolve that are resistant to various fungicides. Biotrophic fungal pathogens colonize living plant tissue and obtain nutrients from living host cells. Necrotrophic fungal pathogens infect and kill host tissue and extract nutrients from the dead host cells. Most bacteria that are associated with plants are actually saprotrophic, and do no harm to the plant itself. However, a small number, around 100 known species, are able to cause disease. Bacterial diseases are much more prevalent in sub-tropical and tropical regions of the world.

Most plant pathogenic bacteria are rod-shaped (bacilli). In order to be able to colonize the plant they have specific pathogenicity factors. Five main types of bacterial pathogenicity factors are known: uses of Cell wall-degrading enzymes, Toxins, Effector proteins, Phytohormones and Exopolysaccharides Pathogens such as *Erwinia*, use Cell wall-degrading enzymes to cause soft rot. *Agrobacterium* changes the level of auxins to cause tumours with phytohormones. Exopolysaccharides are produced by bacteria and block xylem vessels, often leading to the death of the plant. Bacteria control the production of pathogenicity factors via quorum sensing. There are many types of plant virus, and some are even asymptomatic. Under normal circumstances, plant viruses cause only a loss of crop yield. Therefore, it is not economically viable to try to control them, the exception being when they infect perennial species, such as fruit trees. Most plant viruses have small, single-stranded RNA genomes. However some plant viruses also have double stranded RNA or single or double stranded DNA genomes. These genomes may encode only three or four proteins: a replicase, a coat protein, a movement protein, in order to allow cell to cell movement through plasmodesmata, and sometimes a protein that allows transmission by a vector. Plant viruses can have several more proteins and employ many different molecular translation methods. Plant viruses must be transmitted from plant to plant by a vector. This is often by an insect, but some fungi, nematodes, and protozoa have been shown to be viral vectors. In many cases, the insect and virus are specific for virus transmission such as the beet leafhopper that transmits the curly top virus causing disease in several crop plants. Nematodes are small, multicellular wormlike creatures. Many live freely in the soil, but there are some species that parasitize plant roots.

The book fulfil not only the need of the students to find literature on the diseases and other pathological conditions difficult to obtain and access, but also provide complete systematic treatment of the subject from their point of view.

— *Bibhuti Dutta Jha*

Contents

Chapter 1

Viral Diseases of Plants

A virus is a microscopic infectious agent that can reproduce only inside a host cell. Viruses infect all types of organisms: from animals and plants, to bacteria and archaea. Since the initial discovery of tobacco mosaic virus by Martinus Beijerinck in 1898, more than 5,000 types of virus have been described in detail, although most types of virus remain undiscovered. Viruses are ubiquitous, as they are found in almost every ecosystem on Earth, and are the most abundant type of biological entity on the planet. The study of viruses is known as virology, and is a branch of microbiology.

Viruses consist of two or three parts: all viruses have genes made from either DNA or RNA, long molecules that carry genetic information; all have a protein coat that protects these genes; and some have an envelope of fat that surrounds them when they are outside a cell. Viruses vary in shape from simple helical and icosahedral shapes, to more complex structures. They are about 1/100th the size of bacteria. The origins of viruses in the evolutionary history of life are unclear: some may have evolved from plasmids—pieces of DNA that can move between cells—while others may have evolved from bacteria. In evolution, viruses are an important means of horizontal gene transfer, which increases genetic diversity.

Viruses spread in many ways; plant viruses are often transmitted from plant to plant by insects that feed on sap, such as aphids, while animal viruses can be carried by blood-sucking insects. These disease-bearing organisms are known as vectors. Influenza viruses are spread by coughing and sneezing, and others such as norovirus, are transmitted by the faecal-oral route, when they contaminate hands, food, or water. Rotaviruses are often spread by direct contact with infected children. HIV is one of several viruses that are transmitted through sexual contact. Not all viruses cause disease, as many viruses reproduce without causing any obvious harm to the infected organism. Viruses such as hepatitis B can cause life-long or chronic infections, and the viruses continue to replicate in the body despite the hosts' defence mechanisms. In some cases, these chronic infections might be beneficial as they might increase the immune system's response against infection by other pathogens. However, in most cases viral infections in animals cause an immune response that eliminates the infecting virus. These immune responses can also be produced by vaccines that give immunity to a viral infection. Microorganisms such as bacteria also have defences against viral infection, such as restriction modification systems. Antibiotics have no effect on viruses, but antiviral drugs have been developed to treat both life-threatening and more minor infections.

Etymology

The word is from the Latin *virus* referring to poison and other noxious substances, first used in English in 1392. *Virulent*, from Latin *virulentus* (poisonous) dates to 1400. A meaning of "agent that causes infectious disease" is first recorded in 1728, before the discovery of viruses by Dmitry Ivanovsky in 1892. The adjective *viral* dates to 1948. The term *virion* is also used to refer to a single infective viral particle. The plural of virus is "viruses".

History

In 1884, the French microbiologist Charles Chamberland invented a filter (known today as the Chamberland filter or Chamberland-Pasteur filter), with pores smaller than bacteria. Thus, he could pass a solution containing bacteria through the filter and completely remove them from the solution. In 1892 the Russian biologist Dimitri Ivanovski used this filter to study what is now known to be tobacco mosaic virus. His experiments showed that the crushed leaf extracts from infected tobacco plants are still infectious after filtration. Ivanovski suggested the infection might be caused by a toxin produced by bacteria, but did not pursue the idea. At the time it was thought that all infectious agents could be retained by filters and grown on a nutrient medium—this was part of the germ theory of disease. In 1898 the Dutch microbiologist Martinus Beijerinck repeated the experiments and became convinced that this was a new form of infectious agent. He went on to observe that the agent multiplied only in dividing cells, but as his experiments did not show that it was made of particles, he called it a *contagium vivum fluidum* (soluble living germ) and re-introduced the word *virus*. Beijerinck maintained that viruses were liquid in nature, a theory later discredited by Wendell Stanley, who proved they were particulate. In the same year, 1899, Friedrich Loeffler and Frosch passed the agent of foot and mouth disease (aphthovirus) through a similar filter and ruled out the possibility of a toxin because of the high dilution; they concluded that the agent could replicate.

In the early 20th century, the English bacteriologist Frederick Twort discovered the viruses that infect bacteria, which are now called bacteriophages, and the French-Canadian microbiologist Félix d'Herelle described viruses that, when added to bacteria on agar, would produce areas of dead bacteria. He accurately diluted a suspension of these viruses and discovered that the highest dilutions, rather than killing all the bacteria, formed discrete areas of dead organisms. Counting these areas and multiplying by the dilution factor allowed him to calculate the number of viruses in the suspension.

By the end of the nineteenth century, viruses were defined in terms of their infectivity, filterability, and their requirement for living hosts. Viruses had been grown only in plants and animals. In 1906, Harrison invented a method for growing tissue in lymph, and, in 1913, E. Steinhardt, C. Israeli, and R. A. Lambert used this method to grow vaccinia virus in fragments of guinea pig corneal tissue. In 1928, H. B. Maitland and M. C. Maitland grew vaccinia virus in suspensions of minced hens' kidneys. Their method was not widely adopted until the 1950s, when poliovirus was grown on a large scale for vaccine production. Another breakthrough came in 1931, when the American pathologist Ernest William Goodpasture grew influenza and several other viruses in fertilised chickens' eggs. In 1949 John F. Enders, Thomas Weller, and Frederick Robbins grew polio virus in cultured human embryo cells, the first virus to be grown without using solid animal tissue or eggs. This work enabled Jonas Salk to make an effective polio vaccine.

With the invention of electron microscopy in 1931 by the German engineers Ernst Ruska and Max Knoll came the first images of viruses. In 1935 American biochemist and virologist Wendell Stanley examined the Tobacco mosaic virus and found it to be mostly made from protein.

A short time later, this virus was separated into protein and RNA parts. Tobacco mosaic virus was the first one to be crystallised and whose structure could therefore be elucidated in detail. The first X-ray diffraction pictures of the crystallised virus were obtained by Bernal and Fankuchen in 1941. Based on her pictures, Rosalind Franklin discovered the full structure of the virus in 1955.

In the same year, Heinz Fraenkel-Conrat and Robley Williams showed that purified Tobacco mosaic virus RNA and its coat protein can assemble by themselves to form functional viruses, suggesting that this simple mechanism was probably how viruses assembled within their host cells.

The second half of the twentieth century was the golden age of virus discovery and most of the 2,000 recognised species of animal, plant, and bacterial viruses were discovered during these years. In 1957, equine arterivirus and the cause of Bovine virus diarrhea (a pestivirus) were discovered. In 1963, the hepatitis B virus was discovered by Baruch Blumberg, and in 1965, Howard Temin described the first retrovirus. Reverse transcriptase, the key enzyme that retroviruses use to translate their RNA into DNA, was first described in 1970, independently by Howard Temin and David Baltimore. In 1983 Luc Montagnier's team at the Pasteur Institute in France, first isolated the retrovirus now called HIV.

Origins

Viruses are found wherever there is life and have probably existed since living cells first evolved. The origin of viruses is unclear because they do not form fossils, so molecular techniques have been the most useful means of investigating how they arose. These techniques rely on the availability of ancient viral DNA or RNA, but, unfortunately, most of the viruses that have been preserved and stored in laboratories are less than 90 years old. There are three main hypotheses that try to explain the origins of viruses:

Regressive Hypothesis

Viruses may have once been small cells that parasitised larger cells. Over time, genes not required by their parasitism were lost. The bacteria rickettsia and chlamydia are living cells that, like viruses, can reproduce only inside host cells. They lend support to this hypothesis, as their dependence on parasitism is likely to have caused the loss of genes that enabled them to survive outside a cell. This is also called the *degeneracy hypothesis*.

Cellular Origin Hypothesis

Some viruses may have evolved from bits of DNA or RNA that "escaped" from the genes of a larger organism. The escaped DNA could have come from plasmids (pieces of naked DNA that can move *between* cells) or transposons (molecules of DNA that replicate and move around to different positions *within* the genes of the cell). Once called "jumping genes", transposons are examples of mobile genetic elements and could be the origin of some viruses. They were discovered in maize by Barbara McClintock in 1950. This is sometimes called the *vagrancy hypothesis*.

Coevolution Hypothesis

Viruses may have evolved from complex molecules of protein and nucleic acid at the same time as cells first appeared on earth and would have been dependent on cellular life for many millions of years. Viroids are molecules of RNA that are not classified as viruses because they lack a protein coat. However, they have characteristics that are common to several viruses and are often called subviral agents. Viroids are important pathogens of plants. They do not code for proteins but interact with the host cell and use the host machinery for their replication. The hepatitis delta virus of humans has an RNA genome similar to viroids but has protein coat derived from hepatitis B virus and cannot produce one of its own. It is therefore a defective virus and cannot replicate without the help of hepatitis B virus.

The Virophage 'sputnik' infects the Mimivirus and the related Mamavirus, which in turn infect the protozooan *Acanthamoeba castellanii.* These viruses that are dependent on other virus species are called satellites and may represent evolutionary intermediates of viroids and viruses. Prions are infectious protein molecules that do not contain DNA or RNA. They cause an infection in sheep called scrapie and cattle bovine spongiform encephalopathy ("mad cow" disease). In humans they cause kuru and Creutzfeldt-Jakob disease. They are able to replicate because some proteins can exist in two different shapes and the prion changes the normal shape of a host protein into the prion shape. This starts a chain reaction where each prion protein converts many host proteins into more prions, and these new prions then go on to convert even more protein into prions. Although they are fundamentally different from viruses and viroids, their discovery gives credence to the idea that viruses could have evolved from self-replicating molecules.

Computer analysis of viral and host DNA sequences is giving a better understanding of the evolutionary relationships between different viruses and may help identify the ancestors of modern viruses. To date, such analyses have not helped to decide on which of these hypotheses are correct. However, it seems unlikely that all currently known viruses have a common ancestor and viruses have probably arisen numerous times in the past by one or more mechanisms. Opinions differ on whether viruses are a form of life, or organic structures that interact with living organisms. They have been described as "organisms at the edge of life", since they resemble organisms in that they possess genes and evolve by natural selection, and reproduce by creating multiple copies of themselves through self-assembly.

However, although they have genes, they do not have a cellular structure, which is often seen as the basic unit of life. Additionally, viruses do not have their own metabolism, and require a host cell to make new products. They therefore cannot reproduce outside a host cell (although bacterial species such as rickettsia and chlamydia are considered living organisms despite the same limitation). Accepted forms of life use cell division to reproduce, whereas viruses spontaneously assemble within cells, which is analogous to the autonomous growth of crystals. Virus self-assembly within host cells has implications for the study of the origin of life, as it lends further credence to the hypothesis that life could have started as self-assembling organic molecules.

Microbiology

Structure

Viruses display a wide diversity of shapes and sizes, called morphologies. Viruses are about 1/100th the size of bacteria. Most viruses that have been studied have a diameter between 10 and 300 nanometres. Some filoviruses have a total length of up to 1400 nm, however their diameters are only about 80 nm. Most viruses are unable to be seen with a light microscope so scanning and transmission electron microscopes are used to visualise virus particles.

To increase the contrast between viruses and the background, electron-dense "stains" are used. These are solutions of salts of heavy metals such as tungsten, that scatter the electrons from regions covered with the stain. When virus particles are coated with stain (positive staining), fine detail is obscured. Negative staining overcomes this problem by staining the background only.

A complete virus particle, known as a virion, consists of nucleic acid surrounded by a protective coat of protein called a capsid. These are formed from identical protein subunits called capsomers. Viruses can have a lipid "envelope" derived from the host cell membrane. The capsid is made from proteins encoded by the viral genome and its shape serves as the basis for morphological distinction. Virally coded protein subunits will self-assemble to form a capsid, generally requiring the presence of the virus genome.

However, complex viruses code for proteins that assist in the construction of their capsid. Proteins associated with nucleic acid are known as nucleoproteins, and the association of viral capsid proteins with viral nucleic acid is called a nucleocapsid. The capsid and entire virus structure can be mechanically (physically) probed through atomic force microscopy. In general, there are four main morphological virus types:

Helical

These viruses are composed of a single type of capsomer stacked around a central axis to form a helical structure, which may have a central cavity, or hollow tube.

This arrangement results in rod-shaped or filamentous virions: these can be short and highly rigid, or long and very flexible. The genetic material, generally single-stranded RNA, but ssDNA in some cases, is bound into the protein helix, by interactions between the negatively charged nucleic acid and positive charges on the protein.

Overall, the length of a helical capsid is related to the length of the nucleic acid contained within it and the diameter is dependent on the size and arrangement of capsomers. The well-studied Tobacco mosaic virus is an example of a helical virus.

Icosahedral

Most animal viruses are icosahedral or near-spherical with icosahedral symmetry. A regular icosahedron is the optimum way of forming a closed shell from identical sub-units. The minimum number of identical capsomers required is twelve, each composed of five identical sub-units.

Many viruses, such as rotavirus, have more than twelve capsomers and appear spherical but they retain this symmetry. Capsomers at the apices are surrounded by five other capsomers and are called pentons. Capsomers on the triangular faces are surround by six others and are call hexons.

Envelope

Some species of virus envelope themselves in a modified form of one of the cell membranes, either the outer membrane surrounding an infected host cell, or internal membranes such as nuclear membrane or endoplasmic reticulum, thus gaining an outer lipid bilayer known as a viral envelope.

This membrane is studded with proteins coded for by the viral genome and host genome; the lipid membrane itself and any carbohydrates present originate entirely from the host. The influenza virus and HIV use this strategy. Most enveloped viruses are dependent on the envelope for their infectivity.

Complex

These viruses possess a capsid that is neither purely helical, nor purely icosahedral, and that may possess extra structures such as protein tails or a complex outer wall. Some bacteriophages, such as Enterobacteria phage T4 have a complex structure consisting of an icosahedral head bound to a helical tail, which may have a hexagonal base plate with protruding protein tail fibres. This tail structure acts like a molecular syringe, attaching to the bacterial host and then injecting the viral genome into the cell.

The poxviruses are large, complex viruses that have an unusual morphology. The viral genome is associated with proteins within a central disk structure known as a nucleoid. The nucleoid is surrounded by a membrane and two lateral bodies of unknown function. The virus has an outer envelope with a thick layer of protein studded over its surface. The whole particle is slightly pleiomorphic, ranging from ovoid to brick shape. Mimivirus is the largest known virus, with a capsid diameter of 400 nm. Protein filaments measuring 100 nm project from the surface.

The capsid appears hexagonal under an electron microscope, therefore the capsid is probably icosahedral. Some viruses that infect Archaea have complex structures that are unrelated to any other form of virus, with a wide variety of unusual shapes, ranging from spindle-shaped structures, to viruses that resemble hooked rods, teardrops or even bottles. Other archaeal viruses resemble the tailed bacteriophages, although they can have multiple tail structures.

Genome

Genomic diversity among viruses:

Property	*Parameters*
Nucleic acid	• DNA • RNA • Both DNA and RNA (at different stages in the life cycle)
Shape	• Linear • Circular

	• Segmented
Strandedness	• Single-stranded
	• Double-stranded
	• Double-stranded with regions of single-strandedness
Sense	• Positive sense (+)
	• Negative sense (“)
	• Ambisense (+/”)

An enormous variety of genomic structures can be seen among viral species; as a group they contain more structural genomic diversity than plants, animals, archaea, or bacteria. A virus has either DNA or RNA genes and are called DNA viruses and RNA viruses respectively. By far most viruses have RNA. Plant viruses tend to have single-stranded RNA and bacteriophages tend to have double-stranded DNA. Viral genomes are circular, such as polyomaviruses, or linear, such as adenoviruses. The type of nucleic acid is irrelevant to the shape of the genome. Among RNA viruses, the genome is often divided up into separate parts within the virion and is called *segmented.* Each segment often codes for one protein and they are usually found together in one capsid. Every segment is not required to be in the same virion for the overall virus to be infectious, as demonstrated by the brome mosaic virus.

A viral genome, irrespective of nucleic acid type, is either single-stranded or double-stranded. Single-stranded genomes consist of an unpaired nucleic acid, analogous to one-half of a ladder split down the middle. Double-stranded genomes consist of two complementary paired nucleic acids, analogous to a ladder. Some viruses, such as those belonging to the Hepadnaviridae, contain a genome that is partially double-stranded and partially single-stranded.

For viruses with RNA or single-stranded DNA, the strands are said to be either positive-sense (called the plus-strand) or negative-sense (called the minus-strand), depending on whether it is complementary to the viral messenger RNA (mRNA). Positive-sense viral RNA is identical to viral mRNA and thus can be immediately translated by the host cell. Negative-sense viral RNA is complementary to mRNA and thus must be converted to positive-sense RNA by an RNA polymerase before translation. DNA nomenclature is similar to RNA nomenclature, in that the *coding strand* for the viral mRNA is complementary to it ("), and the *non-coding strand* is a copy of it (+).

Genome size varies greatly between species. The smallest viral genomes code for only four proteins and have a mass of about 10^6 Daltons; the largest have a mass of about 10^8 Daltons and code for over one hundred proteins. RNA viruses generally have smaller genome sizes than DNA viruses due to a higher error-rate when replicating, and have a maximum upper size limit. Beyond this limit, errors in the genome when replicating render the virus useless or uncompetitive. To compensate for this, RNA viruses often have segmented genomes where the genome is split into smaller molecules, thus reducing the chance of error. In contrast, DNA viruses generally have larger genomes due to the high fidelity of their replication enzymes.

Viruses undergo genetic change by several mechanisms. These include a process called genetic drift where individual bases in the DNA or RNA mutate to other bases. Most of these point mutations are "silent"—they do not change the protein that the gene encodes—but others can confer evolutionary advantages such as resistance to antiviral drugs. Antigenic shift is where there is a major change in the genome of the virus. This occurs as a result of recombination or reassortment. When this happens with influenza viruses, pandemics may result. RNA viruses often exist as quasispecies or swarms of viruses of the same species but with slightly different genome nucleoside sequences. Such quasispecies are a prime target for natural selection.

Segmented genomes confer evolutionary advantages; different strains of a virus with a segmented genome can shuffle and combine genes and produce progeny viruses or (offspring) that have unique characteristics. This is called reassortment or *viral sex*.

Genetic recombination is the process by which a strand of DNA is broken and then joined to the end of a different DNA molecule. This can occur when viruses infect cells simultaneously and studies of viral evolution have shown that recombination has been rampant in the species studied. Recombination is common to both RNA and DNA viruses.

Replication Cycle

Viral populations do not grow through cell division, because they are acellular; instead, they use the machinery and metabolism of a host cell to produce multiple copies of themselves, and they *assemble* in the cell.

The life cycle of viruses differs greatly between species but there are six *basic* stages in the life cycle of viruses:

- *Attachment* is a specific binding between viral capsid proteins and specific receptors on the host cellular surface. This specificity determines the host range of a virus. For example, HIV infects only human T cells, because its surface protein, gp120, can interact with CD4 and receptors on the T cell's surface. This mechanism has evolved to favour those viruses that only infect cells in which they are capable of replication. Attachment to the receptor can induce the viral-envelope protein to undergo changes that results in the fusion of viral and cellular membranes.
- *Penetration* follows attachment; viruses enter the host cell through receptor mediated endocytosis or membrane fusion. This is often called viral entry. The infection of plant cells is different to that of animal cells. Plants have a rigid cell wall made of cellulose and viruses can only get inside the cells following trauma to the cell wall. Viruses such as tobacco mosaic virus can also move directly in plants, from cell-to-cell, through pores called plasmodesmata. Bacteria, like plants, have strong cell walls that a virus must breach to infect the cell. Some viruses have evolved mechanisms that inject their genome into the bacterial cell while the viral capsid remains outside.
- *Uncoating* is a process in which the viral capsid is degraded by viral enzymes or host enzymes thus releasing the viral genomic nucleic acid.
- *Replication* involves synthesis of viral messenger RNA (mRNA) for viruses except positive sense RNA viruses, viral protein synthesis and assembly of viral proteins and viral genome replication.

- Following the *assembly* of the virus particles, post-translational modification of the viral proteins often occurs. In viruses such as HIV, this modification, (sometimes called maturation), occurs *after* the virus has been released from the host cell.
- Viruses are *released* from the host cell by lysis—a process that kills the cell by bursting its membrane. Enveloped viruses (e.g., HIV) typically are released from the host cell by budding. During this process the virus acquires its envelope, which is a modified piece of the host's plasma membrane.

The genetic material within viruses, and the method by which the material is replicated, vary between different types of viruses.

DNA Viruses

The genome replication of most DNA viruses takes place in the cell's nucleus. If the cell has the appropriate receptor on its surface, these viruses enter the cell by fusion with the cell membrane or by endocytosis. Most DNA viruses are entirely dependent on the host cell's DNA and RNA synthesising machinery, and RNA processing machinery. The viral genome must cross the cell's nuclear membrane to access this machinery.

RNA Viruses

These viruses are unique because their genetic information is encoded in RNA. Replication usually takes place in the cytoplasm. RNA viruses can be placed into about four different groups depending on their modes of replication. The polarity (whether or not it can be used directly to make proteins) of the RNA largely determines the replicative mechanism, and whether the genetic material is single-stranded or double-stranded. RNA viruses use their own RNA replicase enzymes to create copies of their genomes.

Reverse Transcribing Viruses

These replicate using reverse transcription, which is the formation of DNA from an RNA template. Reverse transcribing viruses containing RNA genomes use a DNA intermediate to replicate, whereas those containing DNA genomes use an RNA intermediate during genome replication.

Both types use the reverse transcriptase enzyme to carry out the nucleic acid conversion. Retroviruses often integrate the DNA produced by reverse transcription into the host genome. They are susceptible to antiviral drugs that inhibit the reverse transcriptase enzyme, e.g. zidovudine and lamivudine. An example of the first type is HIV, which is a retrovirus. Examples of the second type are the Hepadnaviridae, which includes Hepatitis B virus.

Effects on the Host Cell

The range of structural and biochemical effects that viruses have on the hosts cell is extensive. These are called *cytopathic effects*. Most virus infections eventually result in the death of the host cell. The causes of death include cell lysis, alterations to the cell's surface membrane and apoptosis. Often cell death is caused by cessation of its normal activities due to suppression by virus-specific proteins, not all of which are components of the virus particle.

Some viruses cause no apparent changes to the infected cell. Cells in which the virus is latent and inactive show few signs of infection and often function normally. This causes persistent infections and the virus is often dormant for many months or years. This is often the case with herpes viruses. Viruses, such as Epstein-Barr virus often cause cells to proliferate without causing malignancy, but viruses, such as papillomaviruses are an established cause of cancer.

Classification

Classification seeks to describe the diversity of viruses by naming and grouping them based on similarities. In 1962, André Lwoff, Robert Horne, and Paul Tournier were the first to develop a means of virus classification, based on the Linnaean hierarchical system.

This system bases classification on phylum, class, order, family, genus, and species. Viruses were grouped according to their shared properties (not of their hosts) and the type of nucleic acid forming their genomes. later the International Committee on Taxonomy of Viruses was formed.

ICTV Classification

The International Committee on Taxonomy of Viruses (ICTV) developed the current classification system and wrote guidelines that put a greater weight on certain virus properties to maintain family uniformity. A universal system for classifying viruses, and a unified taxonomy, has been established since 1966. The 7th lCTV Report formalised for the first time the concept of the virus species as the lowest taxon (group) in a branching hierarchy of viral taxa.

However, at present only a small part of the total diversity of viruses has been studied, with analyses of samples from humans finding that about 20% of the virus sequences recovered have not been seen before, and samples from the environment, such as from seawater and ocean sediments, finding that the large majority of sequences are completely novel.

The general taxonomic structure is as follows:

- Order (-virales)
- Family (-viridae)
- Subfamily (-virinae)
- Genus (*-virus*)
- Species (*-virus*)

In the current (2008) ICTV taxonomy, five orders have been established, the Caudovirales, Herpesvirales, Mononegavirales, Nidovirales, and Picornavirales. The committee does not formally distinguish between subspecies, strains, and isolates. In total there are 5 orders, 82 families, 11 subfamilies, 307 genera, 2,083 species and about 3,000 types yet unclassified.

Baltimore Classification

The Nobel Prize-winning biologist David Baltimore devised the Baltimore classification system. The ICTV classification system is used in conjunction with the Baltimore classification system in modern virus classification.

The Baltimore classification of viruses is based on the mechanism of mRNA production. Viruses must generate mRNAs from their genomes to produce proteins and replicate themselves, but different mechanisms are used to achieve this in each virus family. Viral genomes may be single-stranded (ss) or double-stranded (ds), RNA or DNA, and may or may not use reverse transcriptase (RT). Additionally, ssRNA viruses may be either sense (+) or antisense (-). This classification places viruses into seven groups:

- I: dsDNA viruses (e.g. Adenoviruses, Herpesviruses, Poxviruses)
- II: ssDNA viruses (+)sense DNA (e.g. Parvoviruses)
- III: dsRNA viruses (e.g. Reoviruses)
- IV: (+)ssRNA viruses (+)sense RNA (e.g. Picornaviruses, Togaviruses)
- V: (-)ssRNA viruses (-)sense RNA (e.g. Orthomyxoviruses, Rhabdoviruses)
- VI: ssRNA-RT viruses (+)sense RNA with DNA intermediate in life-cycle
- VII: dsDNA-RT viruses.

As an example of viral classification, the chicken pox virus, varicella zoster (VZV), belongs to the order Herpesvirales, family Herpesviridae, subfamily Alphaherpesvirinae, and genus Varicellovirus. VZV is in Group I of the Baltimore Classification because it is a dsDNA virus that does not use reverse transcriptase.

Viruses and Human Disease

Examples of common human diseases caused by viruses include the common cold, influenza, chickenpox and cold sores. Many serious diseases such as ebola, AIDS, avian influenza and SARS are caused by viruses. The relative ability of viruses to cause disease is described in terms of virulence. Other diseases are under investigation as to whether they too have a virus as the causative agent, such as the possible connection between human herpes virus six (HHV6) and neurological diseases such as multiple sclerosis and chronic fatigue syndrome. There is current controversy over whether the borna virus, previously thought to cause neurological diseases in horses, could be responsible for psychiatric illnesses in humans.

Viruses have different mechanisms by which they produce disease in an organism, which largely depends on the viral species. Mechanisms at the cellular level primarily include cell lysis, the breaking open and subsequent death of the cell. In multicellular organisms, if enough cells die the whole organism will start to suffer the effects. Although viruses cause disruption of healthy homeostasis, resulting in disease, they may exist relatively harmlessly within an organism. An example would include the ability of the herpes simplex virus, which cause cold sores, to remain in a dormant state within the human body. This is called latency and is a characteristic of the all herpes viruses including the Epstein-Barr virus, which causes glandular fever, and the varicella zoster virus, which causes chicken pox. The large majority of people have been infected with at least one of these types of herpes virus. However, these latent viruses might sometimes be beneficial, as the presence of the virus can increase immunity against bacterial pathogens, such as Yersinia pestis. On the other hand, latent chickenpox infections return in later life as the disease called shingles.

Some viruses can cause life-long or chronic infections, where the viruses continue to replicate in the body despite the hosts' defence mechanisms. This is common in hepatitis B virus and hepatitis C virus infections. People chronically infected are known as carriers, as they serve as reservoirs of infectious virus. In populations with a high proportion of carriers, the disease is said to be endemic. In contrast to acute lytic viral infections this persistence implies compatible interactions with the host organism. Persistent viruses may even broaden the evolutionary potential of host species.

Epidemiology

Viral epidemiology is the branch of medical science that deals with the transmission and control of virus infections in humans. Transmission of viruses can be vertical, that is from mother to child, or horizontal, which means from person to person.

Examples of vertical transmission include hepatitis B virus and HIV where the baby is born already infected with the virus. Another, more rare, example is the varicella zoster virus, which although causing relatively mild infections in humans, can be fatal to the foetus and newly born baby. Horizontal transmission is the most common mechanism of spread of viruses in populations. Transmission can be exchange of blood by sexual activity, e.g. HIV, hepatitis B and hepatitis C; by mouth by exchange of saliva, e.g. Epstein-Barr virus, or from contaminated food or water, e.g. norovirus; by breathing in viruses in the form of aerosols, e.g. influenza virus; and by insect vectors such as mosquitoes, e.g. dengue. The rate or speed of transmission of viral infections depends on factors that include population density, the number of susceptible individuals,, the quality of health care and the weather.

Epidemiology is used to break the chain of infection in populations during outbreaks of viral diseases. Control measures are used that are based on knowledge of how the virus is transmitted. It is important to find the source, or sources, of the outbreak and to identify the virus. Once the virus has been identified, the chain of transmission can sometimes be broken by vaccines.

When vaccines are not available sanitation and disinfection can be effective. Often infected people are isolated from the rest of the community and those that have been exposed to the virus placed in quarantine. To control the outbreak of foot and mouth disease in cattle in Britain in 2001, thousands of cattle were slaughtered.

Most viral infections of humans and other animals have incubation periods during which the infection causes no signs or symptoms. Incubation periods for viral diseases range from a few days to weeks but are known for most infections. Somewhat overlapping, but mainly following the incubation period, there is a *period of communicability*; a time when an infected individual or animal is contagious and can infect another person or animal. This too is known for many viral infections and knowledge the length of both periods is important in the control of outbreaks. When outbreaks cause an unusually high proportion of cases in a population, community or region they are called epidemics. If outbreaks spread worldwide they are called pandemics.

Epidemics and Pandemics

Native American populations were devastated by contagious diseases, particularly smallpox, brought to the Americas by European colonists. It is unclear how many Native Americans were killed by foreign diseases after the arrival of

Columbus in the Americas, but the numbers have been estimated to be close to 70% of the indigenous population. The damage done by this disease significantly aided European attempts to displace and conquer the native population.

A pandemic is a worldwide epidemic. The 1918 flu pandemic, commonly referred to as the Spanish flu, was a category 5 influenza pandemic caused by an unusually severe and deadly influenza A virus. The victims were often healthy young adults, in contrast to most influenza outbreaks, which predominantly affect juvenile, elderly, or otherwise weakened patients.

The Spanish flu pandemic lasted from 1918 to 1919. Older estimates say it killed 40–50 million people, while more recent research suggests that it may have killed as many as 100 million people, or 5% of the world's population in 1918. Most researchers believe that HIV originated in sub-Saharan Africa during the twentieth century; it is now a pandemic, with an estimated 38.6 million people now living with the disease worldwide. The Joint United Nations Programme on HIV/AIDS (UNAIDS) and the World Health Organization (WHO) estimate that AIDS has killed more than 25 million people since it was first recognised on June 5, 1981, making it one of the most destructive epidemics in recorded history. In 2007 there were 2.7 million new HIV infections and 2 million HIV-related deaths.

Several highly lethal viral pathogens are members of the Filoviridae. Filoviruses are filament-like viruses that cause viral hemorrhagic fever, and include the ebola and marburg viruses. The Marburg virus attracted widespread press attention in April 2005 for an outbreak in Angola. Beginning in October 2004 and continuing into 2005, the outbreak was the world's worst epidemic of any kind of viral hemorrhagic fever.

Cancer

Viruses are an established cause of cancer in humans and other species. Viral cancers only occur in a minority of infected persons (or animals). Cancer viruses come from a range of virus families, including both RNA and DNA viruses, and so there is no single type of "oncovirus" (an obsolete term originally used for acutely-transforming retroviruses). The development of cancer is determined by a variety of factors such as host immunity and mutations in the host. Viruses accepted to cause human cancers include some genotypes of human papillomavirus, hepatitis B virus, hepatitis C virus, Epstein-Barr virus, Kaposi's sarcoma-associated herpesvirus and human T-lymphotropic virus. The most recently discovered human cancer virus is a polyomavirus (Merkel cell polyomavirus) that causes most cases of a rare form of skin cancer called Merkel cell carcinoma. Hepatitis viruses can develop into a chronic viral infection that leads to liver cancer.

Infection by human T-lymphotropic virus can lead to tropical spastic paraparesis and adult T-cell leukemia. Human papillomaviruses are an established cause of cancers of cervix, skin, anus, and penis. Within the Herpesviridae, Kaposi's sarcoma-associated herpesvirus causes Kaposi's sarcoma and body cavity lymphoma, and Epstein–Barr virus causes Burkitt's lymphoma, Hodgkin's lymphoma, B lymphoproliferative disorder and nasopharyngeal carcinoma. Merkel cell polyomavirus closely related to SV40 and mouse polyomaviruses that have been used as animal models for cancer viruses for over 50 years.

Host Defence Mechanisms

The body's first line of defence against viruses is the innate immune system. This comprises cells and other mechanisms that defend the host from infection in a non-specific manner.

This means that the cells of the innate system recognise, and respond to, pathogens in a generic way, but unlike the adaptive immune system, it does not confer long-lasting or protective immunity to the host. RNA interference is an important innate defence against viruses. Many viruses have a replication strategy that involves double-stranded RNA (dsRNA). When such a virus infects a cell, it releases its RNA molecule or molecules, which immediately bind to a protein complex called dicer that cuts the RNA into smaller pieces.

A biochemical pathway called the RISC complex is activated, which degrades the viral mRNA and the cell survives the infection. Rotaviruses avoid this mechanism by not uncoating fully inside the cell and by releasing newly produced mRNA through pores in the particle's inner capsid. The genomic dsRNA remains protected inside the core of the virion.

When the adaptive immune system of a vertebrate encounters a virus, it produces specific antibodies that bind to the virus and render it non-infectious. This is called humoral immunity. Two types of antibodies are important.

The first called IgM is highly effective at neutralizing viruses but is only produced by the cells of the immune system for a few weeks. The second, called, IgG is produced indefinitely. The presence of IgM in the blood of the host is used to test for acute infection, whereas IgG indicates an infection sometime in the past. IgG antibody is measured when tests for immunity are carried out.

A second defence of vertebrates against viruses is called cell-mediated immunity and involves immune cells known as T cells. The body's cells constantly display short fragments of their proteins on the cell's surface, and if a T cell recognises a suspicious viral fragment there, the host cell is destroyed by *T killer* cells and the virus-specific T-cells proliferate.

Cells such as the macrophage are specialists at this antigen presentation. The production of interferon is an important host defence mechanism. This is a hormone produced by the body when viruses are present. Its role in immunity is complex, but it eventually stops the viruses from reproducing by killing the infected cell and its close neighbours.

Not all virus infections produce a protective immune response in this way. HIV evades the immune system by constantly changing the amino acid sequence of the proteins on the surface of the virion. These persistent viruses evade immune control by sequestration, blockade of antigen presentation, cytokine resistance, evasion of natural killer cell activities, escape from apoptosis, and antigenic shift. Other viruses, called neurotropic viruses, are disseminated by neural spread where the immune system may be unable to reach them.

Prevention and Treatment

Because viruses use the machinery of a host cell to reproduce and reside within them, they are difficult to eliminate without killing the host cell. The most effective medical approaches to viral diseases so far are vaccinations to provide resistance to infection, and antiviral drugs.

Vaccines

Vaccination is a cheap and effective way of preventing infections by viruses. Vaccines were used to prevent viral infections long before the discovery of the actual viruses. Their use has resulted in a dramatic decline in morbidity (illness) and mortality (death) associated with viral infections such as polio, measles, mumps and rubella. Smallpox infections have been eradicated.

Currently vaccines are available to prevent over thirteen viral infections of humans, and more are used to prevent viral infections of animals. Vaccines can consist of live-attenuated or killed viruses, or viral proteins (antigens). Live vaccines contain weakened forms of the virus that causes the disease.

Such viruses are called attenuated. Live vaccines can be dangerous when given to people with a weak immunity, (who are described as immunocompromised), because in these people, the weakened virus can cause the original disease. Biotechnology and genetic engineering techniques are used to produce subunit vaccines. These vaccines use only the capsid proteins of the virus. Hepatitis B vaccine is an example of this type of vaccine. Subunit vaccines are safe for immunocompromised patients because they cannot cause the disease. However, the yellow fever virus vaccine, a live-attenuated strain called 17D, is probably the safest and most effective vaccine ever generated.

Antiviral Drugs

Over the past twenty years, the development of antiviral drugs has increased rapidly. This has been driven by the AIDS pandemic. Antiviral drugs are often nucleoside analogues, (fake DNA building blocks), which viruses incorporate into their genomes during replication. The life-cycle of the virus is then halted because the newly synthesised DNA is inactive. This is because these analogues lack the hydroxyl groups, which, along with phosphorus atoms, link together to form the strong "backbone" of the DNA molecule. This is called DNA chain termination. Examples of nucleoside analogues are aciclovir for Herpes simplex virus infections and lamivudine for HIV and Hepatitis B virus infections. Aciclovir is one of the oldest and most frequently prescribed antiviral drugs. Other antiviral drugs in use target different stages of the viral life cycle. HIV is dependent on a proteolytic enzyme called the HIV-1 protease for it to become fully infectious. There is a large class of drugs called protease inhibitors that inactivate this enzyme. Hepatitis C is caused by an RNA virus. In 80% of people infected, the disease is chronic, and without treatment, they are infected for the remainder of their lives. However, there is now an effective treatment that uses the nucleoside analogue drug ribavirin combined with interferon. The treatment of chronic carriers of the hepatitis B virus by using a similar strategy using lamivudine has been developed.

Infection in other Species

Viruses infect all cellular life and, although viruses occur universally, each cellular species has its own specific range that often infect only that species. Viruses are important pathogens of livestock. Diseases such as Foot and Mouth Disease and bluetongue are caused by viruses. Companion animals such as cats, dogs, and horses, if not vaccinated, are susceptible to serious viral infections. Canine parvovirus is caused by a small DNA virus and infections are often fatal in pups. Like all invertebrates, the honey bee is susceptible to many viral infections. Fortunately, most viruses co-exist harmlessly in their host and cause no signs or symptoms of disease.

Plants

There are many types of plant virus, but often they cause only a loss of yield, and it is not economically viable to try to control them. Plant viruses are often spread from plant to plant by organisms, known as vectors. These are normally insects, but some fungi, nematode worms and single-celled organisms have been shown to be vectors. When control of plant virus infections is considered economical, for perennial fruits for example, efforts are concentrated on killing the vectors and removing alternate hosts such as weeds. Plant viruses are harmless to humans and other animals because they can reproduce only in living plant cells. Plants have elaborate and effective defence mechanisms against viruses. One of the most effective is the presence of so-called resistance (R) genes. Each R gene confers resistance to a particular virus by triggering localised areas of cell death around the infected cell, which can often be seen with the unaided eye as large spots. This stops the infection from spreading. RNA interference is also an effective defence in plants. When they are infected, plants often produce natural disinfectants that kill viruses, such as salicylic acid, nitric oxide, and reactive oxygen molecules.

Bacteria

Bacteriophages are an extremely common and diverse group of viruses. For example, bacteriophages are the most common form of biological entity in aquatic environments, with up to ten times more of these viruses in the oceans than bacteria, reaching levels of 250,000,000 bacteriophages per millilitre of seawater. These viruses infect specific bacteria by binding to surface receptor molecules and then entering the cell. Within a short amount of time, in some cases just minutes, bacterial polymerase starts translating viral mRNA into protein. These proteins go on to become either new virions within the cell, helper proteins, which help assembly of new virions, or proteins involved in cell lysis. Viral enzymes aid in the breakdown of the cell membrane, and, in the case of the T4 phage, in just over twenty minutes after injection over three hundred phages could be released.

The major way bacteria defend themselves from bacteriophages is by producing enzymes that destroy foreign DNA. These enzymes, called restriction endonucleases, cut up the viral DNA that bacteriophages inject into bacterial cells. Bacteria also contain a system that uses CRISPR sequences to retain fragments of the genomes of viruses that the bacteria have come into contact with in the past, which allows them to block the virus's replication through a form of RNA interference. This genetic system provides bacteria with acquired immunity to infection.

Archaea

Some viruses replicate within archaea: these are double-stranded DNA viruses with unusual and sometimes unique shapes. These viruses have been studied in most detail in the thermophilic archaea, particularly the orders Sulfolobales and Thermoproteales. Defences against these viruses may involve RNA interference from repetitive DNA sequences within archaean genomes that are related to the genes of the viruses.

Applications

Life Sciences and Medicine

Viruses are important to the study of molecular and cellular biology as they provide simple systems that can be used to manipulate and investigate the functions of cells. The study and use of viruses have provided valuable information about aspects of cell biology. For example, viruses have been useful in the study of genetics and helped our understanding of the basic mechanisms of molecular genetics, such as DNA replication, transcription, RNA processing, translation, protein transport, and immunology.

Geneticists often use viruses as vectors to introduce genes into cells that they are studying. This is useful for making the cell produce a foreign substance, or to study the effect of introducing a new gene into the genome. In similar fashion, virotherapy uses viruses as vectors to treat various diseases, as they can specifically target cells and DNA. It shows promising use in the treatment of cancer and in gene therapy. Eastern European scientists have used phage therapy as an alternative to antibiotics for some time, and interest in this approach is increasing, due to the high level of antibiotic resistance now found in some pathogenic bacteria.

Materials Science and Nanotechnology

Current trends in nanotechnology promise to make much more versatile use of viruses. From the viewpoint of a materials scientist, viruses can be regarded as organic nanoparticles. Their surface carries specific tools designed to cross the barriers of their host cells. The size and shape of viruses, and the number and nature of the functional groups on their surface, is precisely defined. As such, viruses are commonly used in materials science as scaffolds for covalently linked surface modifications. A particular quality of viruses is that they can be tailored by directed evolution. The powerful techniques developed by life sciences are becoming the basis of engineering approaches towards nanomaterials, opening a wide range of applications far beyond biology and medicine.

Because of their size, shape, and well-defined chemical structures, viruses have been used as templates for organizing materials on the nanoscale. Recent examples include work at the Naval Research Laboratory in Washington, DC, using Cowpea Mosaic Virus (CPMV) particles to amplify signals in DNA microarray based sensors. In this application, the virus particles separate the fluorescent dyes used for signalling in order to prevent the formation of non-fluorescent dimers that act as quenchers. Another example is the use of CPMV as a nanoscale breadboard for molecular electronics.

Weapons

The ability of viruses to cause devastating epidemics in human societies has led to the concern that viruses could be weaponised for biological warfare. Further concern was raised by the successful recreation of the infamous 1918 influenza virus in a laboratory. The smallpox virus devastated numerous societies throughout history before its eradication. There are officially only two centres in the world which keep stocks of smallpox virus-the Russian Vector laboratory, and the United States Centres for Disease Control. But fears that it may be used as a weapon are not totally unfounded; the vaccine for smallpox is not safe — during the years before the eradication of smallpox disease more people became seriously ill as a result of vaccination than did people from smallpox — and smallpox vaccination is no longer universally practiced. Thus, much of the modern human population has almost no established resistance to smallpox.

Virus Structure

At the simplest level, the function of the outer shells (CAPSID) of a virus particle is to protect the fragile nucleic acid genome from:

- Physical damage-Shearing by mechanical forces.
- Chemical damage-UV irradiation (from sunlight) leading to chemical modification.
- Enzymatic damage-Nucleases derived from dead or leaky cells or deliberately secreted by vertebrates as defence against infection.

The protein subunits in a virus capsid are multiply redundant, i.e. present in many copies per particle. Damage to one or more subunits may render that particular subunit non-functional, but does not destroy the infectivity of the whole particle. Furthermore, the outer surface of the virus is responsible for recognition of the host cell. Initially, this takes the form of binding of a specific virus-attachment protein to a cellular receptor molecule. However, the capsid also has a role to play in initiating infection by delivering the genome from its protective shell in a form in which it can interact with the host cell. To form an infectious particle, a virus must overcome two fundamental problems:

1. To assemble the particle utilizing only the information available from the components which make up the particle itself (capsid + genome).
2. Virus particles form regular geometric shapes, even though the proteins from which they are made are irregularly shaped.

How do these simple organisms solve these difficulties? The information to answer these problem lie in the rules of symmetry.

In 1957, Fraenkel-Conrat & Williams showed that when mixtures of purified tobacco mosaic virus (TMV) RNA & coat protein were incubated together, virus particles formed. The discovery that virus particles could form spontaneously from purified subunits without any extraneous information indicated that the particle was in the free energy minimum state & was therefore the favoured structure of the components. This stability is an important feature of the virus particle.

Although some viruses are very fragile & are essentially unable to survive outside the protected host cell environment, many are able to persist for long periods, in some cases for years.

Helical Capsids

Tobacco mosaic virus (TMV) is representative of one of the two major structural classes seen in viruses of all types, those with helical symmetry. The simplest way to arrange multiple, identical protein subunits is to use rotational symmetry & to arrange the irregularly shaped proteins around the circumference of a circle to form a disc.

Multiple discs can then be stacked on top of one another to form a cylinder, with the virus genome coated by the protein shell or contained in the hollow centre of the cylinder. Closer examination of the TMV particle by X-ray crystallography reveals that the structure of the capsid actually consists of a helix rather than a pile of stacked disks.

A helix can be defined mathematically by two parameters:

- amplitude (diameter) &
- pitch (the distance covered by each complete turn of the helix).

Helices are rather simple structures formed by stacking repeated components with a constant relationship (amplitude & pitch) to one another-note that if this simple constraint is broken a spiral forms rather than a helix-unsuitable for containing a virus genome.

TMV particles are rigid, rod-like structures, but some helical viruses demonstrate considerable flexibility & longer helical virus particles are often seen to be curved or bent. Flexibility is important attribute since long helical particles are subject to damage from shear forces & the ability to bend reduces the chance of breakage.

The fact that helical symmetry is a useful way of arranging a single protein subunit to form a particle is confirmed by the large number of different types of virus which have evolved with this capsid arrangement.

Icosahedral (Isometric) Capsids

An alternative way of building a virus capsid is to arrange protein subunits in the form of a hollow quasi-spherical structure, enclosing the genome within. The criteria for arranging subunits on the surface of a solid are more complex than those for building a helix.

In the 1950s, Brenner & Horne (among others) developed sophisticated techniques which enabled them to use electron microscopy to reveal many of the fine details of the structure of virus particles. One of the most useful techniques proved to be the use of electron-dense dyes such as phosphotungstic acid or uranyl acetate to examine virus particles by negative staining. The small metal ions in such dyes are able to penetrate the minute crevices between the protein subunits in a viral capsid to reveal the fine structure of the particle.

Francis Crick & James Watson (1956), were the first to suggest that virus capsids are composed of numerous identical protein sub-units arranged either in helical or cubic (=icosahedral) symmetry. In order to construct a capsid from repeated subunits, a virus must 'know the rules' which dictate how these are arranged. For an icosahedron, the rules are based on the rotational symmetry of the solid, which is known as 2-3-5 symmetry:

- An axis of two-fold rotational symmetry through the centre of each edge
- An axis of three-fold rotational symmetry through the centre of each face
- An axis of five-fold rotational symmetry through the centre of each corner (vertex).

To help you understand this further, you can build your own icosahedral virus particle.

The simplest icosahedral capsids are built up by using 3 identical subunits to form each triangular face, thereby requiring 60 identical subunits to form a complete capsid. A few simple virus particles are constructed in this way, e.g. bacteriophage ØX174:

In most cases, analysis reveals that icosahedral virus capsids contain more than 60 subunits, for the reasons of genetic economy given above. The capsids of picornaviruses provide a good illustration of the construction of icosahedral virus particles.

Enveloped Viruses

'Naked' virus particles, i.e. those in which the capsid proteins are exposed to the external environment are produced from infected cells at the end of the replicative cycle when the cell dies, breaks down & lyses, releasing the virions which have been built up internally. This simple strategy has drawbacks. In some circumstances it is wasteful, resulting in the premature death of the cell. Many viruses have devised strategies to exit the infected cell without its total destruction. This presents a difficulty in that all living cells are covered by a membrane composed of a lipid bilayer. The viability of the cell depends on the integrity of this membrane. Viruses leaving the cell must therefore allow this membrane to remain intact & this is achieved by extrusion (budding) of the particle through the membrane, during which process the particle becomes coated in a lipid envelope derived from the host cell membrane & with a similar composition:

The structure underlying the envelope may be based on helical or icosahedral symmetry & may be formed before or as the virus leaves the cell. In the majority of cases, enveloped viruses use cellular membranes as sites allowing them to direct assembly. The formation of the particle inside the cell, maturation & release are in many cases a continuous process.

The site of assembly varies for different viruses. Not all enveloped viruses bud from the cell surface membrane, many viruses use cytoplasmic membranes such as the golgi complex, others such as herpesviruses which replicate in the nucleus may utilize the nuclear membrane. In these cases, the virus is usually extruded into some form of vacuole, in which it is transported to the cell surface & subsequently released.

Envelope Proteins

If the virus particle became covered in a smooth, unbroken lipid bilayer, this would be its undoing. Such a coating is effectively inert, & though effective as a protective layer preventing desiccation of or enzymatic damage to the particle, would not permit recognition of receptor molecules on the host cell. Therefore, viruses modify their lipid envelopes by the synthesis of several classes of proteins which are associated in one of three ways with the envelope:

- Matrix Proteins: These are internal virion proteins whose function is effectively to link the internal nucleocapsid assembly
- Glycoproteins: These are transmembrane proteins, anchored to the membrane by a hydrophobic domain & can be subdivided into two types, by their function:

o External Glycoproteins-Anchored in the envelope by a single transmembrane domain. Most of the structure of the protein is on the outside of the membrane, with a relatively short internal tail. Often individual monomers associate to form the 'spikes' visible on the surface of many enveloped viruses in the electron microscope. Such proteins are the major antigens of enveloped viruses.

- o Transport Channels-This class of proteins contains multiple hydrophobic transmembrane domains, forming a protein-lined channel through the envelope, which enables the virus to alter the permeability of the membrane, e.g. ion-channels.

Complex Virus Structures

The majority of viruses can be fitted into one of the three structural classes outlined above, i.e. those with helical symmetry, icosahedral symmetry or enveloped viruses based on either of these two. However, there are many viruses whose structure is more complex. In these cases, although the general principles of symmetry already described are often used to build part of the virus shell (this term being appropriate here since such viruses often consist of several layers of protein & lipid), the larger & more complex viruses cannot be simply defined by a mathematical equation as can a simple helix or icosahedron. Because of the complexity of some of these viruses, they have defied attempts to determine detailed atomic structures using the techniques described earlier.

Poxviruses

An example of such a group & the problems of complexity is shown by the members of the poxvirus family. These viruses have oval or 'brick-shaped' particles 200-400nm long. In fact, these particles are so large that they were first observed using high resolution optical microscopes in 1886 & thought at that time to be 'the spores of micrococci'. The external surface of the virion is ridged in parallel rows, sometimes arranged helically. The particles are extremely complex & have been shown to contain more than 100 different proteins. Antigenically, poxviruses are very complex, inducing both specific & cross-reacting antibodies-hence ability to vaccinate against one disease with another virus.

Poxviruses & a number of other complex viruses also emphasise the true complexity of some virus-there are at least ten enzymes present in poxvirus particles, mostly concerned with nucleic acid metabolism/genome replication.

Tailed Phages

In other cases, the particle structure of complex viruses has been much more completely investigated. One of the prime examples of such a group are the tailed phages of Escherichia coli.

These viruses have been extensively studied for excellent reasons-they are easy to propagate in bacterial cells & can be obtained in high titres & are easily purified, facilitating biochemical & structural studies.

The head of the particle consists essentially of an icosahedral shell attached via a collar to a contractile, helical tail. At the end of the tail is a plate which functions in attachment to the bacterial host & also in penetration of the bacterial cell wall by virtue of lysozyme-like enzymes associated with the plate.

In addition to these structures, thin protein fibres are attached to the plate, which along with the tail plate, are involved in binding to the receptor molecules in the wall of the host cell. In detail, the structure of these phages is rather more complex than this simple picture, for example, there are several internal proteins & polyamines associated with the genomic DNA in the head & an internal tube structure inside the outer sheath of the helical tail.

In the infected bacterial cell, there are separate assembly pathways for the head & tail sections of the particle, which come together at a late stage to make up the virion.

Protein-nucleic Acid Interactions & Genome Packaging

In most cases, the linear virus genome when stretched out in solution is at least an order of magnitude longer than the diameter of the capsid.

Folding the genome in order to stuff it into such a confined space is quite a feat of topology, but is compounded by repulsion by the cumulative negative electrostatic charges on the phosphate groups of the nucleotide backbone resulting in the genome resisting being crammed into a small space. Viruses overcome this difficulty by packaging along with the genome a number of positively-charged molecules in order to counteract this negative charge repulsion. These include:

- small positively charged ions (Na, Mg, K, etc)
- polyamines
- various nucleic acid-binding proteins.

Some of these latter proteins are virus-encoded & contain amino acids with basic side-chains such as arginine & lysine which interact with the genome.

Many viruses with double-stranded DNA genomes have basic histone-like molecules closely associated with the DNA. Some of these are virus-encoded, in other cases, the virus may utilize cellular proteins, e.g. the polyomavirus genome assumes a chromatin-like structure in association with four cellular histone proteins, similar to that of the host cell genome.

The second problem the virus must overcome is to achieve the specificity required to select & encapsidate the virus genome from the large background of cellular nucleic acids. In most cases, by the late stages of virus infection when assembly of virus particles occurs, transcription of cellular genes has been reduced & a large pool of virus genomes have accumulated. The overproduction of viral nucleic acids eases but does not eliminate the problem of specific genome packaging. Therefore, a specific virus-encoded capsid or nucleocapsid protein is required to achieve this end. Hence many viruses, even those with relatively short, compact genomes such as retroviruses & rhabdoviruses encode this type of protein.

The other side of the packaging equation are the specific nucleotide sequences in the genome (the packaging signal) which permit the virus to select genomic nucleic acids from the cellular background. The packaging signal from a number of virus genomes has been identified.

Viruses with segmented genomes face further problems. Not only must they encapsidate only viral nucleic acid & exclude host cell molecules, but they must also attempt to package one of each of the required genome segments. Like many other aspects of virus assembly, the way in which packaging is controlled is not well understood, but the key must lie in the specific molecular interactions between the genome & the capsid.

In general, the physical structure of virus genomes within virus particles has been poorly studied.

Virus Receptors-Recognition & Binding

Cellular receptor molecules used by a number of different viruses from diverse taxonomic groups have now been identified. This is achieved by binding of a specific virus-attachment protein on the outer surface of the virion to a cellular receptor molecule on the outer surface of the host cell. The interaction of viruses with a cellular receptors is a major event in determining the subsequent events in replication & the outcome of infections. It is this interaction which 'activates' extracellular virus particles & initiates the replication cycle.

Other Interactions of the Virus Capsid with the Host Cell

The function of the virus capsid is not only to protect the genome, but also to deliver it to a suitable host cell & more specifically, the appropriate compartment of the host cell (in the case of eukaryote hosts) to allow replication to proceed.

An example are the nucleocapsid proteins of viruses which replicate in the nucleus of the host cell. These molecules contain within their primary amino acid sequences 'nuclear localization signals' which are responsible for the migration of the virus genome plus its associated proteins into the nucleus where replication can occur.

Virions are not inert structures. Many virus particles contain one or more enzymatic activities, although in most cases these are usually not active outside the biochemical environment of the host cell. All viruses with negative-sense RNA genomes must carry with them a virus-specific RNA-dependent RNA polymerase since uninfected cells have no mechanism for RNA-dependent RNA polymerization & therefore, genome replication could not occur if this enzyme were not included in the virus particle. The more complex DNA viruses carry a multiplicity of enzymes in the particles, mostly concerned with nucleic acid metabolism in some way.

An Introduction to the Structure of Viruses

Scientists have long sought to uncover the structure and function of viruses. Viruses are unique — they have been classified as both living and nonliving at various points in the history of biology. What makes them so interesting? Let's look at the structure of viruses.

Structure

A virus particle, also known as a virion, is essentially a nucleic acid (DNA or RNA) enclosed in a protein shell or coat. Viruses are extremely small, approximately 15-25 nanometers in diameter.

Genetic Material

Viruses may have double-stranded DNA, double-stranded RNA, single-stranded DNA or single-stranded RNA.

In different viruses, which of the four is the "genetic material," depends on the nature and function of the specific virus. The viral genome can consist of a very small number of genes or up to hundreds of genes depending on the type of virus. Note that the genome is typically organized as a long molecule that is usually straight or circular.

Protein Coat

The protein coat that envelopes the genetic material is known as a capsid. It can have several shapes: polyhedral, rod or "complex." The protein subunits of the capsid are called capsomeres. In addition to the protein coat, some viruses have specialized structures. For example, the flu virus has a membrane-like envelope around its capsid. The envelope has both host cell and viral components and assists the virus in infecting its host.Capsid additions are also found in bacteriophages. For example, bacteriophages can have a protein "tail" attached to the capsid that is used to infect the host bacteria.

Share Your Opinions

What do you think? How does the structure of a virus help determine its function? Would you classify a virus as living or nonliving? Come on over to the Biology Forum and share your thoughts, opinions and feelings.

How do Viruses Reproduce?

In the first part of this series about viruses, we looked at virus structure and function. As is true in most organisms, structure and function are thoroughly intertwined. This week we look at a fascinating subject — virus replication.

Replication

A single virus particle (virion) is in and of itself essentially inert. It lacks needed components that cells have to reproduce. Viruses are intracellular obligate parasites which means that they cannot reproduce or express their genes without the help of a living cell.

Once a virus has "infected" a cell, it will "marshal" the cell's ribosomes, enzymes and much of the cellular machinery to reproduce. Unlike what we have seen in mitosis and meiosis, viral reproduction produces many, many progeny, that when complete, leave the host cell to infect other cells in the organism. The exact nature of what happens after the host is infected varies depending on the nature of the virus. In most cases, the process depends on the form of the genome. The process for double-stranded DNA, single-stranded DNA, double-stranded RNA and single-stranded RNA will differ.

Self-Assembly

Interestingly enough, once the viral progeny components are produced by the cellular machinery, the assembly of the viral genome and the viral capsids is a non-enzymatic process. It is usually spontaneous.

Specificity

Ah, the beauties of structure and function! Viruses typically can only infect a limited number of hosts (also known as host range). The "lock and key" mechanism is the most common explanation for this range. Certain proteins on the virus particle must fit certain receptor sites on the particular host's cell surface.

Chapter 2

Non-pesticide Diseases

Pests and Parasites

Varroa Mites

Varroa destructor and *Varroa jacobsoni* are parasitic mites that feed off the bodily fluids of adult, pupal and larval bees. *Varroa* mites can be seen with the naked eye as a small red or brown spot on the bee's thorax. *Varroa* are carriers for a virus that is particularly damaging to the bees. Bees that are infected with this virus during their development will often have visibly deformed wings. *Varroa* have led to the virtual elimination of feral bee colonies in many areas and is a major problem for kept bees in apiaries. Some feral populations are now recovering — it appears that they have been naturally selected for *Varroa* resistance (these so-called feral populations may be africanized bees).

Varroa were first discovered in Southeast Asia in about 1904, but is now present on all continents except Australia. *Varroa* were discovered in the United States in 1987, in New Zealand in 2000, and in the United Kingdom in 1992Devon).

Varroa are generally not a problem for a hive that is growing strongly. When the hive population growth reduced in preparation for winter or due to poor late summer forage the mite population growth can overtake that of the bees and can then destroy the hive. Often a colony will simply *abscond* under such conditions. *Varroa* in combination with Deformed Wing Virus and bacteria have been theoretically implicated in Colony Collapse Disorder.

Treatment

A variety of treatments are currently marketed or practiced to attempt to control Varroa mites. The treatments are generally segregated into chemical controls and "mechanical" controls.

Common chemical controls include "hard" chemicals such as fluvalinate and coumaphos and "soft" chemicals such as thymol, sucrose octanoate esters, oxalic acid and formic acid (sold in gel packs "Mite-Away" but also used in other formulations). According to the U.S. Environmental Protection Agency, when used in beehives as directed, these treatments kill a large proportion of the mites while not substantially disrupting bee behaviour or life span. Use of chemical controls is generally regulated and varies from country to country. With few exceptions, they are not intended for use during production of marketable honey.

Common "mechanical" controls generally rely on disruption of some aspect of the mites' lifecycle. These controls are generally intended not to eliminate all mites but merely to maintain the infestation at a level which the colony can tolerate. Examples of mechanical controls include drone brood sacrifice (varroa mites are preferentially attracted to the drone brood), powdered sugar dusting (which encourages cleaning behaviour and dislodges some mites), screened bottom boards (so that any dislodged mites fall through the bottom and away from the colony), brood interruption and, perhaps, downsizing of the brood cell size. A device called the Varroa Mite Control Entrance (VMCE) is under development as of 2008. The VMCE works in conjunction with a screened bottom board, by dislodging varroa mites from bees as they enter and exit a hive.

Acarine (Tracheal) Mites

Acarapis woodi is a small parasitic mite that infests the airways of the honey bee. The first known infestation of the mites occurred in the British Isles in the early 20th century. First observed on the Isle of Wight in 1904, the mystery illness known as *Isle of Wight Disease* was not identified as being caused by a parasite until 1921. It quickly spread to the rest of Great Britain. It was regarded as having wiped out the entire bee population of the British Isles (although later genetic studies have found remnants that did survive) and it dealt a devastating blow to British beekeeping. Brother Adam at the Buckfast Abbey developed a resistant hybrid bee known as the Buckfast bee, which is now available worldwide to combat acarine disease.

Diagnosis for tracheal mites generally involves the dissection and microscopic examination of a sample of bees from the hive.

Acarine mites, formerly known as tracheal mites are believed to have entered the US in 1984, *via* Mexico.

Mature female acarine mites leave the bee's airway and climb out on a hair of the bee, where they wait until they can transfer to a young bee. Once on the new bee, they will move into the airways and begin laying eggs.

Treatment

Acarine mites are commonly controlled with grease patties (typically made from 1 part vegetable shortening mixed with 3-4 parts powdered sugar) placed on the top bars of the hive. The bees come to eat the sugar and pick up traces of shortening, which disrupts the mite's ability to identify a young bee. Some of the mites waiting to transfer to a new host will remain on the original host. Others will transfer to a random bee-a proportion of which will die of other causes before the mite can reproduce. Menthol, either allowed to vaporize from crystal form or mixed into the grease patties, is also often used to treat acarine mites.

Nosema

Nosema apis is a spore-forming parasite that invades the intestinal tracts of adult bees and causes nosema disease. Nosema is also associated with Black queen-cell virus. Nosema is normally only a problem when the bees can not leave the hive to eliminate waste (for example, during an extended cold spell in winter or when the hives are enclosed in a wintering barn). When the bees are unable to void (cleansing flights), they can develop dysentery. Nosema is treated by increasing the ventilation through the hive. Some beekeepers will treat a hive with antibiotics. Nosema can also be prevented or minimized by removing much of the honey from the beehive then feeding the bees on sugar water in the late fall. Sugar water made from refined sugar has lower ash content than flower nectar, reducing the risk of dysentery, and may have essentially the same nutritional content, although this remains a point of controversy among some beekeepers. In 1996, a similar type of organism to *Nosema apis* was discovered on the Asian honey bee Apis cerana and subsequently named *Nosema ceranae*. This parasite apparently also infects the Western honey bee.

It has been reported that exposure to corn pollen containing genes for *Bacillus thuringiensis* (Bt) production may weaken the bees' defence against *Nosema*. In this study, it is stated that in relation to feeding a group of bees with Bt corn pollen and a control group with non-Bt corn pollen, that: "in the first year the bee colonies happened to be infested with parasites (microsporidia). This infestation led to a reduction in the number of bees and subsequently to reduced broods in the Bt-fed colonies as well as in the colonies fed on Bt-toxin-free pollen. The trial was therefore discontinued at an early stage. This effect was significantly more marked in the Bt-fed colonies. (The significant differences indicate an interaction of toxin and pathogen on the epithelial cells of the honeybee intestine. The underlying mechanism which causes this effect is unknown.)"

Small Hive Beetle

Aethina tumida is a small, dark-coloured beetle that lives in beehives.

Originally from Africa, the first discovery of small hive beetles in the western hemisphere occurred in the US. The first identified specimen was found in St. Lucie, FL in 1998. The earliest specimens confirmed since then were collected from Charleston, SC in 1996. By December 1999, small hive beetle was reported in Iowa, Maine, Massachusetts, Minnesota, New Jersey, Ohio, Pennsylvania, Texas, and Wisconsin, and was found in California by 2006.

The life cycle of this beetle includes pupation in the ground outside of the hive. Controls to prevent ants from climbing into the hive are believed to also be effective against the hive beetle. Several beekeepers are experimenting with the use of diatomaceous earth around the hive as a way to disrupt the beetle's lifecycle. The diatoms abrade the insect's surface, causing them to dehydrate and die.

Several pesticides are currently used against the small hive beetle. The chemical is commonly applied inside the corrugations of a piece of cardboard. Standard corrugations are large enough that a small hive beetle will enter the cardboard through the end but small enough that honey bees can not enter (and thus are kept away from the pesticide). Alternative controls (such as cooking-oil-based bottom board traps) are also becoming available. Also available are beetle eaters that go between the frames that uses cooking oil.

Wax Moths

Galleria mellonella (greater wax moths) will not attack the bees directly, but feed on the wax used by the bees to build their honeycomb. Their full development to adults requires access to used brood comb or brood cell cleanings — these contain protein essential for the larvae's development, in the form of brood coocoons.

The destruction of the comb will spill or contaminate stored honey and may kill bee larvae.

When honey supers are stored for the winter in a mild climate, or in heated storage, the wax moth larvae can destroy portions of the comb, even though they will not fully develop. Damaged comb may be scraped out and will be replaced by the bees. Wax moth larvae and eggs are killed by freezing, so storage in unheated sheds or barns in higher latitudes is the only control necessary.

Because wax moths cannot survive a cold winter, they are usually not a problem for beekeepers in the northern U.S. or Canada, unless they survive winter in heated storage, or are brought from the south by purchase or migration of beekeepers. They thrive and spread most rapidly with temperatures above 30°C (90°F), so some areas with only occasional days that hot, rarely have a problem with wax moths, unless the colony is already weak due to stress from other factors.

Control and Treatment

A strong hive generally needs no treatment to control wax moths; the bees themselves will kill and clean out the moth larvae and webs. Wax moth larvae may fully develop in cell cleanings when such cleanings accumulate thickly where they are not accessible to the bees. Wax moth development in comb is generally not a problem with top bar hives as unused combs are usually left in the hive during the winter. Since this type of hive is not used in severe wintering conditions, the bees will be able to patrol and inspect the unused comb.

Wax moths can be controlled in stored comb by application of the aizawai variety of Bt (Bacillus thuringiensis) spores via spraying. It is a very effective biological control and has an excellent safety record.

Wax moths can be controlled chemically with paradichlorobenzene (moth crystals or urinal disks). If chemical methods are used, the combs must be well-aired-out for several days before use. The use of naphthalene (mothballs) is discouraged because it accumulates in the wax, which can kill bees or contaminate honey stores. Control of wax moths by other means includes the freezing of the comb for at least twenty-four hours.

Bacterial Diseases

American Foul Brood (AFB), caused by the spore-forming *Paenibacillus larvae ssp. larvae* (formerly classified as *Bacillus larvae*), is the most widespread and destructive of the bee brood diseases. Paenibacillus larvae is a rod-shaped bacterium, which is visible only under a high power microscope. Larvae up to 3 days old become infected by ingesting spores that are present in their food. Young larvae less than 24 hours old are most susceptible to infection. Spores germinate in the gut of the larva and the vegetative form of the bacteria begins to grow, taking its nourishment from the larva. Spores will not germinate in larvae over 3 days old. Infected larvae normally die after their cell is sealed. The vegetative form of the bacterium will die but not before it produces many millions of spores. Each dead larva may contain as many as 100 million spores. This disease only affects the bee larvae but is highly infectious and deadly to bee brood. Infected larvae darken and die.

History

Until 1906 the two foulbrood diseases were not differentiated and the condition was generally referred to as foulbrood. Phillips (1906) used the terms European and

American to distinguish the diseases. However the designations did not refer to the geographical distributions but to the areas where they were first investigated scientifically. White (1907) demonstrated conclusively that a bacterium that he called Bacillus larvae was the cause of American Foulbrood (AFB) disease by fulfilling Koch's postulates. The geographical origin of AFB is unknown, but it is found almost worldwide.

Diagnosis

Lab testing is necessary for definitive diagnosis, but a good field test is to touch a dead larva with a toothpick or twig. It will be sticky and "ropey" (drawn out). Foulbrood also has a characteristic odour, and experienced beekeepers with a good sense of smell can often detect the disease upon opening a hive. In the photo at right, some larvae are healthy while others are diseased. Capped cells with decomposing larvae are sunken, as can be seen at lower right. Some caps may be torn, as well. Compare with healthy brood. The most reliable disease diagnosis is done by sending in some possibly affected brood comb to a laboratory specialized in identifying honey bee diseases.

Disease Spread

When cleaning infected cells, bees distribute spores throughout the entire colony. Disease spreads rapidly throughout the hive as the bees, attempting to remove the spore-laden dead larvae, contaminate brood food. Nectar stored in contaminated cells will contain spores and soon the brood chamber becomes filled with contaminated honey. As this honey is moved up into the supers, the entire hive becomes contaminated with spores. When the colony becomes weak from AFB infection, robber bees may enter and take contaminated honey back to their hives thereby spreading the disease to other colonies and apiaries.

Beekeepers also may spread disease by moving equipment (frames or supers) from contaminated hives to healthy ones. American Foul Brood spores are extremely resistant to desiccation and can remain viable for more than 40 years in honey and beekeeping equipment. Therefore honey from an unknown source should never be used as bee feed, and used beekeeping equipment should be assumed contaminated unless known to be otherwise.

Treatment

AFB spores are present in virtually every hive. Some brood in weakened colonies can become diseased. If the diseased larva dies within the hive, millions of spores are released. Antibiotics, in non-resistant strains of the pathogen, can prevent the vegetative state of the bacterium forming. Drug treatment to prevent the American foulbrood spores from successfully germinating and proliferating is possible using oxytetracycline hydrochloride (Terramycin). Another drug treatment is tylosin tartrate that was US Food and Drug Administration (FDA) approved in 2005.

Chemical treatment is sometimes used prophylactically, but this is a source of considerable controversy because certain strains of the bacterium seem to be rapidly developing resistance. In addition, hives that are contaminated with millions of American foulbrood spores have to be prophylactically treated indefinitely. Once the treatment is suspended the American foulbrood spores germinate successfully again leading to a disease outbreak. Because of the persistence of the spores (which can survive up to 40 years), many State Apiary Inspectors require an AFB diseased hive to be burned completely. A less radical method of containing the spread of disease is burning the frames and comb and thoroughly flame scorching the interior of the hive body, bottom board and covers. Dipping the hive parts in hot paraffin wax or a 3% sodium hypochlorite solution (bleach) also renders the AFB spores innocuous.

European Foulbrood (EFB)

Melissococcus plutonius is a bacterium that infests the mid-gut of an infected bee larva. European foulbrood is less deadly to a colony than American foulbrood. *Melissococcus plutonius* does not form spores, though it can overwinter on comb. European foulbrood is often considered a "stress" disease-a disease that is dangerous only if the colony is already under stress for other reasons. An otherwise healthy colony can usually survive European foulbrood. An outbreak of the disease may be controlled chemically with oxytetracycline hydrochloride, but honies from treated colonies could have chemical residues from the treatment. The 'Shook Swarm' technique of bee husbandry can also be used to effectively control the disease, the advantage being that chemicals are not used. Prophylactic treatments are not recommended as they lead to resistant bacteria.

Fungal Diseases

Chalkbrood

Ascosphaera apis is a fungal disease that infests the gut of the larva. The fungus will compete with the larva for food, ultimately causing it to starve. The fungus will then go on to consume the rest of the larva's body, causing it to appear white and 'chalky'. Chalkbrood is most commonly visible during wet springs. Hives with Chalkbrood can generally be recovered by increasing the ventilation through the hive and/or by requeening the hive.

Stonebrood

Stonebrood is a fungal disease caused by *Aspergillus fumigatus, Aspergillus flavus* and *Aspergillus niger*. It causes mummification of the brood of a honey bee colony. The fungi are common soil inhabitants and are also pathogenic to other insects, birds and mammals. The disease is difficult to identify in the early stages of infection. The spores of the different species have different colours and can also cause respiratory damage to humans and other animals. When a bee larva takes in spores they may hatch in the gut, growing rapidly to form a collarlike ring near the head. After death the larvae turn black and become difficult to crush, hence the name stonebrood. Eventually the fungus erupts from the integument of the larva and forms a false skin. In this stage the larvae are covered with powdery fungal spores. Worker bees clean out the infected brood and the hive may recover depending on factors such as the strength of the colony, the level of infection, and hygienic habits of the strain of bees (there is variation in the trait among different subspecies/races).

Viral Diseases

Acute Bee Paralysis Virus (ABPV) or (APV)

ABPV is considered to be a common infective agent of bees. It belongs to the family Dicistroviridae, as does the Israel acute paralysis virus, Kashmir bee virus, and the Black queen cell virus. It is frequently detected in apparently healthy colonies. Apparently, this virus plays a role in cases of sudden collapse of honey bee colonies infested with the parasitic mite *Varroa destructor*.

Israel Acute Paralysis Virus (IAPV)

A related virus described in 2004 is known as the "Israel acute paralysis virus"; The virus is named after the place where it was first idenitified-its place of origin is unknown. IAPV has been suggested in September 2007 as a marker associated with Colony Collapse Disorder.

Kashmir Bee Virus (KBV)

This is another Dicistroviridae, related to the preceding viruses. Recently discovered, KBV is currently only positively identifiable by a laboratory test. Little is known about it yet.

Black Queen Cell Virus (BQCV)

This is another Dicistroviridae, related to the preceding viruses. As its name implies, BQCV causes the queen larva to turn black and die. It is thought to be associated with Nosema.

- Chronic Paralysis Virus [CPV]
- Cloudy Wing Virus (CWV)
- Deformed Wing Virus (DWV).

As indicated by the name, this virus produces deformed wings. Typically associated with *Varroa destructor*, it has been suggested as a contributing factor to Colony Collapse Disorder.

Sacbrood Virus (SBV)

Morator aetatulas is the virus that causes sacbrood disease. Affected larvae change from pearly white to gray and finally black. Death occurs when the larvae are upright, just before pupation.

Consequently, affected larvae are usually found in capped cells. Head development of diseased larvae is typically retarded. The head region is usually darker than the rest of the body and may lean toward the centre of the cell.

When affected larvae are carefully removed from their cells, they appear to be a sac filled with water. Typically the scales are brittle but easy to remove. Sacbrood-diseased larvae have no characteristic odour.

Dysentery

Dysentery is a condition resulting from a combination of long periods of inability to make *cleansing flights* (generally due to cold weather) and food stores which contain a high proportion of indigestible matter. As a bee's gut becomes engorged with feces that cannot be voided in flight as preferred by the bees, the bee voids within the hive. When enough bees do this the hive population rapidly collapses and death of the colony results. Dark honeys and honeydews have greater quantities of indigestible matter. Occasional warm days in winter are critical for honey bee survival; dysentery problems increase in likelihood if there are periods of more than two or three weeks with temperatures below 50 degrees Fahrenheit. When cleansing flights are few, bees will often be forced out at times when the temperature is barely adequate for their wing muscles to function, and large quantities of bees may be seen dead in the snow around the hives.

Colonies that are found dead in spring from dysentery will have feces smeared over the frames and other hive parts. In very cold areas of North America and Europe, where honey bees are kept in ventilated buildings during the coldest part of winter, no cleansing flights are possible; under such circumstances, it is common for beekeepers to remove all honey from the hives and replace it with sugar water or high fructose corn syrup, which have nearly no indigestible matter.

Chilled Brood

Chilled brood is not actually a disease but can be a result of mistreatment of the bees by the beekeeper. It also can be caused by a pesticide hit that primarily kills off the adult population, or by a sudden drop in temperature during rapid spring buildup. The brood must be kept warm at all times; nurse bees will cluster over the brood to keep it at the right temperature. When a beekeeper opens the hive (to inspect, remove honey, check the queen, or just to look) and prevents the nurse bees from clustering on the frame for too long, the brood can become chilled, deforming or even killing some of the bees.

To minimize the risk of chilled brood, open the hive on warm days and at the hottest part of the day (this is also the time when the most field bees will be out foraging and the number of bees in the hive will be at its lowest). Learn to inspect your hive as quickly as possible and put frames with brood back where the bees can cluster on it immediately.

Pesticide Losses

Honey bees are susceptible to many of the chemicals used for agricultural spraying of other insects and pests. Many pesticides are known to be toxic to bees. Because the bees forage up to several miles from the hive, they may fly into areas actively being sprayed by farmers or they may collect pollen from 'contaminated' flowers.

Carbamate pesticides, such as Sevin(R)-Carbaryl (C12H11NO2)can be especially pernicious since toxicity can take as long as two days to become effective; allowing infected pollen to be returned and distributed throughout the colony. Organophosphates and other insecticides are also known to kill honey bee clusters in treated areas.

Pesticide losses may be relatively easy to identify (large and sudden numbers of dead bees in front of the hive) or quite difficult, especially if the loss results from a gradual accumulation of pesticide brought in by the foraging bees. Quick acting pesticides may deprive the hive of its foragers, dropping them in the field before they can return home.

Insecticides that are toxic to bees have label directions that protect the bees from poisoning as they forage. To comply with the label, applicators must know where and when bees forage in the application area, and the length of residual activity of the pesticide. Some pesticide authorities recommend, and some jurisdictions require, that notice of spraying be sent to all known beekeepers in the area so that they can seal the entrances to their hives and keep the bees inside until the pesticide has had a chance to disperse. This, however, does not solve all problems associated with spraying and the label instructions should be followed regardless of doing this. Sealing honey bees from flight on hot days can kill bees. Beekeeper notification does not offer any protection to bees, if the beekeeper cannot access them, or to wild native or feral honey bees. Thus beekeeper notification as the sole protection procedure does not really protect all the pollinators of the area, and is, in effect, a circumventing of the label requirements. Pesticide losses are a major factor in pollinator decline.

Colony Collapse Disorder

Colony Collapse Disorder (or CCD) is a little-understood phenomenon in which worker bees from a beehive or Western honey bee colony abruptly disappear. CCD was originally found in Western honey bee colonies in North America in late 2006.

European beekeepers observed a similar phenomenon in Belgium, France, the Netherlands, Greece, Italy, Portugal, and Spain, and initial reports have also come in from Switzerland and Germany, albeit to a lesser degree. Possible cases of CCD have also been reported in Taiwan since April 2007. The cause (or causes) of the syndrome is not yet well understood. Hypotheses include environmental change-related stresses, malnutrition, pathogens (i.e., disease including Israel acute paralysis virus), mites, pesticides such as neonicotinoids or imidacloprid, radiation from cellular phones or other man-made devices, and genetically modified (GM) crops with pest control characteristics such as transgenic maize. Some claim that the disappearances have not been reported from organic beekeepers, suggesting to some that beekeeping practices can be a primary factor.

Methods to Assess Exposure to Pesticides

Modelling Techniques to Estimate Pesticide Exposures

For many years, information on the distribution and fate of chemical substances was acquired largely through retrospective studies involving large-scale monitoring that incorporated an enormous number of chemical analyses. The retrospective approach leaves wide margins for error in the environmental management of chemical substances, for it provides a weak base from which to anticipate problems.†

One approach to reduce this difficulty is to forecast the environmental behaviour of a pesticide through the use of computerized models. Over the past decade, some investigators have advanced the idea that one can predict the behaviour of a chemical on the basis of comparisons of measured properties with those of compounds for which greater amounts of environmental data are available. For example, the concept was introduced for the use of hypothetical 'evaluative models' based on properties of stylized environments and pollutants to provide quantitative approaches to exposure estimation.

Models and Modelling why use Them?

Environmental scientists must assess impacts of pesticides in a multiplicity of situations. It is not practical to establish the associated exposure profiles through direct measurement except in rare cases. Thus, environmental scientists and pesticide regulators require tools to estimate exposures to pesticides from one situation to another.

Tools are needed first to optimize the design of monitoring programs, and secondly to test the validity of (a) study designs used in existing monitoring programs, and (b) compilations of data to estimate environmental pollutant levels. A valid study design must include the sampling of at least the matrices in which the pesticide is most likely to be found at any given time. In complex systems, which matrix, stratum, or sampling site is most likely to be contaminated is not necessarily obvious. Models provide the analyst with the tools to optimize the design of monitoring schemes and hence to increase the chance of detecting a compound, should it be present.

Such tools are predictive models that depict how a pesticide may behave in various environments. These models may vary in complexity from structured conceptual models of a pesticide's behaviour in a specified situation to complex mathematical integrations that attempt to describe various processes of translocation and removal.

The descriptions of processes can be simple, empirically determined relationships; or they can be complex, theoretically-based relationships, reflecting a chemical's properties and their influence on its translocation and degradation. The exposure models currently used are composed of (a) simple ones that characterize a single input, transport, and removal process controlling the fate of a chemical, and (b) complex models of numerous, simultaneous environmental influences on the ultimate distribution of a chemical. The usefulness of a model is dependent on its ability to provide correlations between the nature of a chemical and its mobility and persistence in a given matrix. A hypothetical model of the environment, sometimes called the 'environmental system' is first constructed by conceptualizing several environmental 'compartments' considered to have distinct, uniform characteristics, such as pH, percentage of organic matter, or temperature. The discharge, loading, or emission rates of a chemical are then described by discrete models. Finally, mathematical relationships are developed to describe the intercompartmental transfer and removal processes that influence the dynamics of a chemical.

The first group of models provides general qualitative descriptions of a chemical in fixed scenarios. They are used as basic screens for ranking pesticides according to their affinity for various ecological compartments, as well as for the evaluation of data available for understanding the distribution, translocation, and transformation processes. The second group of models has flexible mathematical bases not fixed to specific scenarios. That is, the user can adjust a scenario to develop a comparative base of exposure profiles. Conceivably, they can also be used as a first approach to predict field distribution and planning studies.

The third group, the site-specific models, are the most difficult to develop, for they are directed at providing detailed descriptions of a pesticide's behaviour in a specific complex ecosystem. The number of processes and compartments in such models requires the use of complex approaches such as sensitivity analysis. Development of such models requires that the ecosystem be described in detail and that measures be available to define the transfer and degradation processes operative within various compartments. This modelling is normally achieved through specific empirical measurement of the rates. The conclusions resulting from the use of such models are being questioned at present, and their usefulness is questioned by some. At least two reasons exist for this scepticism. First, some managers and scientists have extended conclusions beyond the limits of the assumptions and errors inherent in the models. Second, modellers have added complexity to models without fully evaluating either the implications of the limitations of the input data or the assumptions used to construct the models. Both groups of individuals seem to forget that models are evaluative tools that are useful only within the boundaries of their limitations in their output. The first two types of models are relevant in hazard assessment schemes, while the third is required in the estimation of risks or benefits.

Validation of Environmental Fate and Distribution Models

The validation of a model and the modelling process must be performed at different levels depending on the type of model and its ultimate use. In levels 1 and 2, qualitative results are sought; hence, the accuracy of trends identified by the models is the focus of attention rather than the numerical values themselves. At level 3, the convergence between the quantitative output of the models and actual field exposure patterns must be examined.

In either case, two types of error must be examined: first, that resulting from the use of incorrect model structure and assumptions; and secondly that resulting from inaccuracies of the input data.

In levels 1 and 2 qualitative models, trend analyses may be generated from correlations of the sorptive and partition properties of the chemicals with their chemical properties. Generally, correlations based on empirically determined characteristics of chemicals are relatively imprecise, particularly when chemicals with widely differing properties are compared. These correlations have been found to be so imprecise as to fail to characterize trends between isomers. This finding may be of particular concern in hazard assessment at level 2, for a chemical's toxicity can be highly dependent on its isomeric structure.

Correlations based on theoretically calculated properties of a molecule (e.g., fragmentation constants) may provide a sound basis for developing qualitative submodels. The use of non-empirical correlations provides improved internal consistency for comparisons, thereby leading to increased accuracy to predict patterns and trends between chemicals of differing structure. In the qualitative models, the most useful trends are generated when sorptive and advective, rather than degradative, processes dominate, because of the paucity of knowledge about relationships among degradative biological processes.

None of the level 3 models are truly validated. Several models can show convergence between field studies and their output; however, arbitrary decisions are generally made about the characteristics of an ecosystem, a chemical's properties, and the models. Hence, the apparent convergence does not constitute validation. Laboratory studies can be of considerable value in improving the accuracy of correlations between the physical-chemical parameters and partition-sorption, translocation, and kinetic data. The accuracy of the input data and the models should be based on the objective of the modelling. Simulation chambers and small-scale pilot experiments have a role in demonstrating the validity of trends predicted by the levels 1 and 2 models. An iterative process exists for the validation of models between their refinement and testing in simulation chambers, small-scale experiments, and field trials. Refinement will occur only through an interactive process of testing and adjustment.

Usefulness of Models to Predict Exposure Profiles

The most accurate models address equilibrium soil-sorption and fish bioconcentration relationships. Models of the sorption of pesticides by other components of aquatic systems are, at best, site-specific; and general predictive relationships are limited. The dynamics under non-equilibrium conditions and at high rates of degradation are at best site-specific and empirically determined. Thus, accurate quantitative predictions are unlikely, given the level of sophistication of today's general models. With sufficient study, site-specific models of a given situation can be developed in which empirical descriptions of the processes are utilized. The degree of precision of such models make them useful for projecting trends within the system, but the results generally cannot be extrapolated to dissimilar situations. A reason for the limitations is the lack of focus on descriptions of models of the processes. Only as these descriptions improve will models provide useful generic qualitative predictions of exposure.

Value of Models to Design Monitoring Programmes and to Evaluate Data

The design of an optimal monitoring program requires the ability to predict the distribution and fate of a chemical at various times within the area to be monitored. Conceptual models of expected equilibrium distribution patterns and dynamic models are both of value.

Those models are particularly useful prior to loss or degradation of significant quantities of a chemical, for it is during this period that simple distribution processes are most likely to govern chemical distribution. In such cases, the accuracy of the evaluations is directly related to that of the description of the relationships between a chemical's partitioning in various phases of an ecosystem. The correlations can be relatively precise and simple.

The predictive relationships for pesticide distribution patterns have sufficient accuracy to be generally useful when the descriptions of the ecosystems are accurate. The difficulty is that accurate models are difficult to obtain in large, complex systems such as rivers and ground-water aquifers. For example, the migration rate of a pesticide through a clay matrix is dependent on understanding the degree of fracturing, a parameter that can only be obtained after detailed geological studies. In the case of a river, good hydrological models are required to predict the location of matrices most likely to contain a pesticide at a given time after its introduction. Where such models exist, they can be very useful to optimize the design of sample programs.

Methods to Measure Exposure to Pesticides

The sophistication of current analytical and separation techniques makes the acquisition of exposure data relatively simple. Unfortunately in many cases, this has resulted in data collection being an end in itself rather than a means to an end. Some people view monitoring as simply the collection and storage of data; actually, it is much more. The United Nations defines monitoring as 'the collection of analytical data according to an agreed plan of sampling in time and space'. Thus, monitoring data are gathered in response to a defined need or objective, established in consultation between data users and analysts. The definition implies that data should be generated in the most appropriate units to ensure comparability and that a sampling program should be constructed prior to analysis by users of the data.

Consequently, data should be as relevant as possible, and the number of analyses performed should be the minimum compatible with the intended purposes. Because these criteria have not yet been employed, much of the data gathered in the past regrettably cannot be used by modellers, regulators, or others.

Uses of Monitoring Data

The three principal users of monitoring data are:

1. Regulators wanting measurements of chemicals in various media either to determine regulatory compliance or to predict concentrations and environmental behaviour,
2. Environmental scientists seeking knowledge of concentrations in environmental compartments either to validate environmental transport models or to describe environmental processes, and
3. Risk analysts characterizing risks to NTOs from the use of pesticides.

Purposes of Monitoring Data

Data obtained from well-designed monitoring programs can be used for numerous purposes in the fields of public health and environmental protection. For example, these objectives include determinations of trends in environmental concentrations, persistence of bioaccumulation of residues, environmental transport and fate, and estimation of doses to NTOs. Furthermore, monitoring data should be an integral part of validating predictive models.

These data must be collected at appropriate time intervals and in correct environmental compartments to answer authoritatively questions raised by users of this information; otherwise, the measurements of pesticide levels at one point in time in various environmental compartments may be of little value to characterize the dose regimen or to validate the models. Monitoring pesticide concentrations in various environmental compartments (i.e., food, air, biota, water) is required to obtain baseline values and trends over time, thus providing information from which to identify emerging problems. Most pesticide monitoring programmes analyse only the parent active ingredient, thus failing to take into account any environmental transformation products or other potentially toxic compounds used in pesticide formulations. Either group of substances may be not only persistent but also of toxicological importance. Such information would help identify NTOs at risk, thereby characterizing more comprehensively the overall risk posed by pesticide formulations. Such data would also provide information on environmental persistence and transport, thereby having intrinsic value to validate predictive models.

The mobility, persistence, and distribution of pesticides (e.g., herbicides contaminating ground or surface waters) in the environment can be ascertained through adequate monitoring.

The establishment of causal relationship uncovered in environmental epidemiology studies depend on appropriate definitions of dosage. However, to date, environmental epidemiological studies have suffered from a lack of such information. For example, determining exposure from measurements at one point in time causes much uncertainty in the determination of dose. Accurate assessment of integrated dosage requires monitoring programs designed from inputs obtained from epidemiologists and chemists. Such measurements need to be frequent and extensive, with samples collected from relevant locations. Such information can, in turn, be used to refine the design of epidemiological studies.

Fate and Transformation: Human

As pesticidal NTOs of considerable public health concern, humans have come under considerable study to ascertain how and to what extent they are exposed to pesticidal substances. Monitoring human exposures has most often taken the form of measuring the presence of the substance of interest in one or more environmental media: air, food, water, or consumer products.

Monitoring has taken place not only at home and at play but also at work. Recently, such monitoring has focused on characterizing human doses most closely associated with injurious consequences, namely, the biological or tissue dose. By using all reasonably accessible tissues, a relatively precise definition of body burden can be obtained, and can aid in deciding appropriate occupational and environmental policies.

The Concept of Exposure Monitoring

The term 'exposure' refers to the quantity of a foreign chemical entering the organism for a specified duration. Depending on the monitoring methods, other terms may prove useful. 'External exposure' may be expressed as concentrations in inhaled air, intake from food and water, or contamination of the skin surface. 'Internal dose' is the amount that enters the circulation, and may be computed using the respective transfer coefficients for retention in the lungs, uptake in the gastrointestinal tract, or absorption through the skin.

With a continuous flux of a foreign chemical into the organism, the term 'dose rate' represents a satisfactory measure of exposure, irrespective of the route by which the contact occurred; the term is also applicable when two or more routes of entry occur. In real-life situations, however, the information is often insufficient for such computations, and fractional data from each route are used. The 'internal dose' or 'dose rate' may be measured as concentration(s) of the chemical and its metabolites in tissues and body fluids, including blood and urine. In principle, the procedure, called 'biological monitoring', represents the simplest way by which dose rates received by various routes can be integrated, depicting total exposure. A link between the data obtained from biological monitoring and those of external environmental monitoring is often secured by studying respective correlations obtained under field conditions. Such investigations are possible, however, with only one route of entry. For multiple exposure routes, individual transfer coefficients are needed. Other important factors to interpret biological monitoring data in relation to external exposure are duration of exposure and time interval after termination of exposure. Chemicals in general and pesticides in particular differ dramatically in their toxicokinetic properties, which are manifest as the presence or absence of accumulation in the body during repeated and prolonged exposure, and as slow versus rapid disappearance following cessation of exposure. Biological monitoring may be based on the measurement of either the unchanged substance or selected metabolites in a particular biological medium. For instance, DDT exposure may be measured as 'total DDT' in blood or fat tissue (biopsy or autopsy) e.g., during occupational high exposureor as the ultimate metabolite, DDA, in urine. In practice, any such determination may be useful, with the method chosen based on availability.

Contemporary trends in biological monitoring of organic chemicals focus on the products of microsomal metabolism. As a result of activation in the system of cytochrome P_{450}-dependent oxygenating enzymes, several organic chemicals yield 'active metabolites' such as epoxides or oxidation products of N-hydroxyderivatives of aromatic compounds.

These active metabolites may bind covalently with macromolecules, such as nucleic acids or proteins, resulting in relatively stable 'adducts' that may be measured selectively. Two separate lines emerge in these trends, of which one centres on adducts with DNA (measured in the white blood cells) and the other on adducts with haemoglobin (measured in the red blood cells). This type of monitoring requires sophisticated methodologies such as P-post-labelling linked with thin layer chromatography or mass spectrometry of the hydrolysis products of the adducts. Therefore, the development of such methods is largely confined to a few advanced and well-equipped laboratories. For pesticides, the best known example of such monitoring has been the fumigant ethylene oxide, which yields DNA adducts.

One important property of macromolecular adducts is their relative stability. For instance, adducts to haemoglobin persist in a red blood cell throughout its life-span; therefore, the readings are likely to increase greatly with the duration of exposure. This situation offers a convenient way of integrating individual doses received over time, yielding the average daily exposure level.

Patterns of Exposure

In practical terms, the collection of information relevant to human exposure to pesticides can be assigned to two broad categories: environmental (or external) measurements such as air, water, and diet; and biological (or internal) measurements. Biological measurements can be conveniently divided into actual measures of the pesticide (or metabolites) or physiological variables. The latter are surrogate measures that must have an established causal relationship with a chemical of interest to be relevant to exposure. In many situations, pesticide exposure results from multiple sources. In an agricultural setting, for example, workers may receive dermal and inhalation doses from pesticide spray application. In the home, exposure can occur through ingestion of unwashed garden produce that contains pesticide residues and from inhalation of indoor pest control sprays. Dietary exposure to pesticides occurs for almost all members of the population. Depending upon the use pattern, one or two exposure routes may be of dominant importance and account for a majority of exposure. Nevertheless, secondary or relatively minor routes may also be influential in the summation of total exposure. Designs for exposure studies must consider statistically correct sampling schemes, and include sufficient numbers of samples to account for loss or attrition, control or baseline measures, and potentially small but significant variations in exposure patterns. To study cholinesterase inhibition, it is frequently important to have baseline or control samples for comparison with those from potentially exposed individuals. Multiple samples are useful to establish ranges of biological variability and to reduce uncertainties surrounding the biological significance of measurements. Many blood variables exhibit cyclic and diurnal patterns; multiple samples taken through time help to distinguish among pesticide exposure-induced changes and normal variability.

Ultimately, total exposure is the relevant measure for assessment. Integration of all routes of exposure, along with characterization of the frequency and duration of exposure, leads to a complete exposure assessment. Knowledge of the endpoint or effects of exposure is important to structure study designs that are statistically strong and relevant to the nature of the exposure. At present, many methodologies are still primarily focused upon single-route exposures. To some extent, an approximation of total exposure is calculated via summation of exposure routes. While this technique is a good start toward total assessment, better understanding of the ways to integrate different routes of exposure is needed. Growing use of carefully designed monitoring and assessment studies which utilize repeated sampling regimes, and which are sensitive to small variations in patterns, is one way to promote improved dosimetry.

Specimen Selection

A variety of human tissues and excreta can be sampled for determination of exposure to pesticides.

Chemical Characteristics

The chemical characteristics of a xenobiotic such as a pesticide are of primary importance to determine how the compound and its metabolites are distributed in and excreted by the body. A major characteristic is lipophilicity (hydrophobicity), which to a large extent governs distribution in adipose tissue. Lipophilic compounds tend to be stored in fatty tissues, and are not excreted into the urine unless metabolized to a relatively water soluble form (e.g., glucuronides). Excretion of fat soluble compounds may occur through the bile duct, but in most instances such compounds are reabsorbed through the enterohepatic circulation. Accordingly, highly lipophilic compounds that are also poorly metabolized, such as DDT, may persist in the body for many years, and may be measurable only through sampling of adipose tissue; however, water soluble compounds or metabolites may be immediately detectable in urine. Other important chemical characteristics include the presence of specific chemical moieties capable of being acted upon by the human metabolic machinery. In some cases, metabolism can produce a specific detectable product in blood or urine; in others, it can act too rapidly and completely degrade a xenobiotic, so that it cannot be readily detected in biological media. Varying rates of metabolism resulting from differing genetic or environmental backgrounds can produce wide variations in the relationship between measured metabolites and external dose, thereby complicating exposure assessment.

The choice of environmental or personal monitoring approaches often depends on technical feasibility. Protecting the worker by measuring a chemical in exposure media is particularly problematic for pesticide applicators and agricultural workers, because exposures usually occur outdoors.

Thus, the standard approach in an enclosed workplace of using air measurements is relatively impractical. Furthermore, worker exposure to pesticides is often through skin contact, which is difficult to measure under any circumstance. These problems, as well as the technical problems related to human monitoring, are detailed elsewhere in this document.

Ethical Aspects of Human Monitoring Sampling Approaches

Some sampling approaches are far more feasible than others in terms of technical difficulty. However, a major overall principle in selecting a human monitoring approach is the absolute need to do no harm to the person undergoing the procedure. This principle restricts the sampling media to excreta and to those tissues that can be sampled with little or no risk to the subject.

In the case of blood, venipuncture is considered to be a small risk, as long as proper sterile techniques are followed and the amount of blood withdrawn is appropriate for the age, size, and health status of the individual. If analytical procedures permit, use of capillary blood obtained by finger puncture should carry less risk. Willingness to undergo human monitoring often depends on perceived risk, discomfort, or inconvenience of the procedure. Implicit in this discussion is that any sampling of an individual must be to that individual's direct benefit. Fully informed consent must always be obtained.

Multi-media and Multi-pesticide Considerations

Pesticide exposures do not always occur through one medium. A single field pesticide application may result in a body burden through (a) inhalation during spray application, (b) passage through the skin resulting from direct skin contact or contact with treated soil, (c) ingestion in food or water consumed by field workers, and (d) ingestion of contaminated soil particularly by children.

Exposure to multiple pesticides occurs when workers use mixtures of pesticides on different crops, and when the consumer ingests food purchased at the supermarket or uses pesticides while gardening as a hobby.

These different, and sometimes unexpected, routes of single or multiple exposures must be considered when selecting specimens for study. Furthermore, human monitoring is necessarily the required approach for significant multiple exposures.

Composition of Pesticide Products

Specimen selection should also consider the various inerts, solvents, and other ingredients in pesticide formulations, since they are almost always a composite of several chemicals. Pesticide formulations may contain the following:

1. One or several chemicals that possess the desired pesticidal property, i.e., chemical 'A' for destruction of broad-leafed plants and herbicide 'B' for grassy plants;
2. Chemicals that enhance the pesticidal properties, such as surfactants or stabilizers;
3. 'Trace contaminants' contained in the base chemicals used in the synthesis of the pesticide or formed during the synthetic process, and
4. Solvents or carriers that facilitate the application of the pesticide; these materials often constitute 9099 per cent of the total volume of pesticide.

Solvents may possess toxic properties that are of human or ecotoxicity concern. Perchloroethylene is an example of the former, toluene one of the latter. Trace contaminants may include nitrosamines and chlorinated dibenzo-*p*-dioxins.

Regional Differences

A number of cultural aspects determine specimen selection. In certain parts of the world, obtaining a blood specimen is particularly difficult because of local beliefs about the importance of blood. In less-developed countries, there might be regions in which it is not technically feasible to obtain, handle, or store certain types of specimens. There are also areas of the world in which legal and administrative burdens are sufficiently imposing as to hinder certain types of human monitoring.

Specific Media

The following media are addressed briefly to indicate their respective merits for biological monitoring: urine, hair, blood, adipose tissue, and breast milk.

Urine

Urine is a useful medium for human monitoring of water soluble pesticides or metabolites. An example is chlordimeform, which can be monitored through measurement of its urinary metabolite, chloro-*o*-toluidine. Several countries require such monitoring for workers using chlordimeform. The profile of urinary metabolites of a single pesticide can sometimes be helpful in distinguishing the pattern or time of exposure. The advantages of using human urine to determine exposure include its ready availability, the non-invasiveness of sampling, and, in some countries, reduced legal or cultural impediments. A major disadvantage is the need for refrigerated storage, to prevent bacterial contamination. The extent and timing of urine collections will often depend on the exposure pattern and toxicokinetics of a pesticide. The use of human urine as a means of measuring exposure to pesticides should be encouraged increasingly. Enhanced understanding of the toxicokinetics of environmental chemicals in recent years permits increased confidence in extrapolation of measurement in various body compartments to the initial dose. This knowledge has been accompanied by rapid advances in analytical chemistry to detect ever smaller amounts of parent compounds and their metabolites.

Hair

Hair is a useful matrix to detect the presence of metals and metaloids such as mercury and arsenic, each of which has been an active ingredient of some pesticides. Hair can be obtained easily and non-invasively. Furthermore, one can take advantage of its relatively slow growth to differentiate from chronic exposure by analysing different hair segments. Nevertheless, a number of problems exist with the use of hair for the determination of exposure. Foremost is the need to distinguish between substances that have been incorporated into the hair matrix during its formation and those that have been added to the hair as part of grooming preparations or through the deposition of dust. Great care must be taken to clean and remove added substances, a process that can be technically difficult.

Blood

Monitoring humans exposed to pesticides is frequently performed through blood sampling. Both direct measurement of the pesticide or its metabolites, and indirect approaches through the measurement of a biological effect of a pesticide (e.g., acetylcholinesterase (AChE) activity) are used.

Blood is a tissue containing cellular and plasma components that can be assayed separately or together. It contains components that will attract both hydrophilic and hydrophobic substances. Thus, not only will water soluble compounds travel readily in the plasma, but highly lipophilic substances can be accommodated in such hydrophobic matrices as the red blood cell membrane, serum albumin, and lipoproteins.

Recent interest in biomarkers has led to the study of potential macromolecular adducts of environmental chemicals using human white blood cells as a source of DNA and albumin or red cell haemoglobin as a source of protein. Haemoglobin is of particular interest, because it is present in very large quantities in routine blood samples, is readily purified, and has a normal lifetime in the blood of approximately 120 days, thereby allowing it to be used to integrate exposure over 4 months and, by density-red-cell-age-separation techniques, theoretically permitting detection and dating of a brief exposure. By contrast, the term 'biological monitoring' refers to the measurement of an effect in the indicator medium, especially in the case of organophosphate pesticides. So far, the only effect utilized has been the inhibition of the AChE in the red blood cells. This measurement is a special example where the effect of real concern (inhibition of the AChE at nerve synapses) is reflected also in the inhibition of the enzyme contained in the blood, which serves as the indicator medium.

When interpreting the results of such an 'effect monitoring', one has to distinguish it clearly from data on 'chemical monitoring'. Both phenomena differ in their toxicokinetics; therefore, a simple relation can be found only in the steady state of the AChE activity. The inhibition of AChE activity occurs to varying degrees in response to equal doses of varous OP pesticides. Therefore, this method is used, for practical purposes, as a self standing (i.e., independent) measure of exposure.

Adipose Tissue

Analysis of body fat samples is a valuable approach to characterize the body burden of lipophilic compounds with relatively long half-lives in the body, such as chlorinated hydrocarbon pesticides. While occasionally used for diagnostic purposes or as part of a research protocol, fat biopsy is too invasive a procedure to be acceptable for routine monitoring of work populations exposed to pesticides.

Adipose tissue banks have been found to be particularly useful for trend analysis, such as following the body burden of DDT, DDE, and PCBs over time when regulatory controls were imposed.

These banks, for the most part, depend on obtaining a fat sample during an operative procedure or autopsy. Archiving tissues is also valuable for future use as new analytical techniques become available. The sampling strategy must be responsive to the specific aims of the fat bank; i.e., for trend analysis, one needs a sampling approach that considers age, sex, health status, and occupation, and one that does not change from year to year. As with any tissue bank, archival integrity is as crucial as the storage process. To be used successfully, adipose tissue banks need to combine archival expertise with the ability to develop and test relevant hypotheses.

Breast Milk

Various lipophilic substances have been shown to be eliminated from the body in breast milk, thereby posing a potential health risk to the nursing infant. Breast milk concentrates body stores of lipophilic substances, acting to displace these stores from the mother to the infant who, with much less total body fat, may well have a higher fat concentration of the unwanted substance. In essence, monitoring of breast milk permits estimation of pesticide exposure to both mother and infant. Water soluble pesticides might also appear in breast milk, but the mobilization and concentration phenomena for lipophilic substances would not be anticipated.

Nursing mothers may be exposed to pesticides in the food they eat or through occupational exposures that occur before or during pregnancy and lactation. In many of the lesser developed countries and in migratory farm workers in developed countries, mothers with babies may actively participate in agricultural efforts including pesticide use; thus, the infant in the field may be at risk both from direct exposure and, if the pesticide is lipid soluble, from breast milk as well. Presently, techniques to obtain breast milk are relatively straightforward, although care must be taken in storage of samples. The usual approach is to relate the concentration of the chemical to milk fat content rather than to total volume. Unfortunately, analytical techniques to measure fat soluble substances, including pesticides, in mother's milk are now technically difficult and costly. A high priority should be given to development of monitoring techniques useful for the detection of pesticides in the breast milk of nursing mothers engaged in agriculture.

The value of assessing breast milk in exposure monitoring for pesticides should be considered in the context of the overall benefit of breast feeding for infant development.

Occupational Exposure

Occupational exposure to agricultural chemicals is generally recognized as having the potential for high doses to the individual, because of the magnitude and frequency of contact. Thus, the number of acute illnesses among agricultural workers reported to national health agencies is a significant public health matter.

The basic approach to estimate occupational exposure is passive dosimetry, biological monitoring, or a combination of the two approaches. Passive techniques are generally used to satisfy regulatory requirements rather than as a means of personal exposure monitoring. As such, these techniques are useful to establish a likely dose under a defined set of field conditions; however, they are not commonly used to monitor exposures of workers on a continuing basis.

Biological monitoring approaches have been discussed fully in previous sections. These techniques may be applied to evaluate occupational exposures as a means of establishing doses of a compound; however, the manner or route of exposure cannot be established by this method alone. The newer immunoassay methodologies hold promise for use as personal monitoring devices. One approach utilizes a small card that, using Enzyme-Linked ImmunoSorbent Assay (ELISA) techniques, can measure pesticide concentrations in virtually any medium. The general approach includes having an individual carry a card containing appropriate antisera and reagents; after exposure to the pesticide, the card changes colour; the colour may then be compared to that on a chart to estimate the concentration of pesticide in the sample. Such cards have already been developed for parathion and paraquat, two chemicals which account for a significant number of acute poisonings in the field. This approach could be used to quickly monitor residues on skin or in urine, and hence to estimate doses to individuals in the field. This approach may prove especially useful in developing nations where access to analytical equipment may be limited.

Occupational exposures to pesticides can be reduced through protective clothing, engineering controls, and modification/refinement of application techniques. Absorption through the skin, in particular the hands, is the major route of exposure for most pesticides. Depending on the specific pesticide and occupational activity, estimates of exposure through the skin of an unprotected hand range from 25 to 98 per cent of total dermal exposure.

The use of protective gloves would greatly reduce occupational pesticide exposure; however, care must be taken to select a glove that is impermeable to the pesticide. For pesticides that are volatile or form aerosols, the use of respirators may be necessary to reduce exposure. However, these devices are limited by the requirement for periodic maintenance and a lack of compliance among workers because of discomfort and heat. In tropical climates, full protective gear may be impractical because of the potential for heat stress in the worker. In such instances, a reduction in exposure can be achieved only through engineering controls and improved application techniques. The use of pre-weighed packages reduces exposure by eliminating the need to handle large volumes of pesticide while weighing out or measuring the amounts necessary for application. In addition, the potential for spills is reduced. The use of emulsifiable concentrates rather than wettable powders reduces inhalation exposure, as does pumping rather than pouring the pesticide from the mixing tank into an application apparatus. The use of closed-cab ground application systems, rather than airplane or airblast, can also reduce exposure to the applicator and reduce drift onto nontarget areas. When airplane applications are used, mechanical rather than human flaggers should be employed to direct the spraying.

Since plants carry a negative electrical charge, the use of an electrostatic device to create an airblast mist that carries a positive electrical charge can cause an electrical attraction between the pesticide mist and plants. This attraction minimizes drift, improves application efficiency and coverage, and ultimately reduces the volume of pesticide required to achieve an effect. Reducing worker exposures through engineering controls and improved application techniques also leads to a reduction in the overall environmental impact of the pesticide application specifically, less drift, fewer spills, and a reduced potential for accidental exposure of other unprotected individuals. Improved techniques for personal monitoring, such as the currently evolving immunoassay methodologies, are desirable to ensure the efficacy and safety of engineering controls and application techniques.

Non-occupational Exposure

In some instances, pesticides used in and around the house can result in meaningful exposure. The practice of treating houses for the control of structural pests has resulted in persistent elevated exposure to the cyclodiene pesticides aldrin/dieldrin and to chlordane/heptachlor.

Periodic application of insecticides for control of vermin is a potential source of exposure to persistent chlorinated hydrocarbon and other products such as chlorpyrifos. Small children may receive the more significant levels of exposure as a consequence of repeated contact with room perimeters, carpets, and soil. In addition, hand to mouth manipulations are common at this age, as is the tendency to taste, chew, and ingest objects. Children of all ages also have the potential for more frequent exposure to pesticides in the home yard, particularly treated lawns or treated soil.

The casual use of pesticides by home owners on the lawn, in the flower garden, or in the vegetable plot, must be considered in determining sources of pesticide exposure. Failure to heed instructions for use of protective clothing, preparation of correct concentrations, and proper disposal can result in levels of exposure equivalent to those present in large-scale agricultural applications. Such lax practices may also result in increased exposure residues on foods harvested from the home garden.

Dietary Exposure

An ideal system to estimate dietary exposures to pesticides should consist of two main components: sufficient data on food consumption that accurately reflect the diverse eating habits and patterns present in a population and its subgroups; and accurate data on the concentration of pesticide residues present in food at the time it is eaten.

Food consumption data used to derive dietary exposure estimates should reflect the diverse eating habits of a population. In some cases, such diversity can be accounted for by using safety factors or exaggerated estimates of residues in food. In such cases, the use of a single population average for food consumption in combination with exaggerated residue estimates will provide adequate protection for all individuals.

The result of using this approach is to greatly overestimate dietary exposures. In other cases, however, it is desirable to obtain the most accurate estimate of exposure possible because of concern for potential toxicity in a particular population subgroup. For example, infants and children are known to consume more food per kilogram of body weight than do adults. Using a population average for food consumption will always under-estimate consumption in children and generally over-estimate consumption in adults. In addition, infants and young children tend to receive a greater portion of their total diet from a small number of foods than do adults.

This restriction of the diet to a few foods, mainly milk products and certain vegetables and fruits, can also magnify pesticide exposures in infants and children, since, if these foods have been treated with a pesticide, a greater portion of the infant diet will contain pesticide residues than that of an adult. Since children are known to be uniquely sensitive to some forms of toxicity (e.g., lead neuropathy), it is important to have accurate consumption data for this group to assess the impact of pesticide residues in their diet. Similarly, for reproductive toxicants, accurate data on food consumption must be obtained for women of child-bearing age.

The toxicity produced by a pesticide can be caused by either a single exposure or multiple exposures over time. Aldicarb is a carbamate insecticide that can produce toxicity after a single exposure, as has been demonstrated by incidents involving the misuse of this chemical on watermelons and cucumbers. To protect against this type of occurrence, it is necessary to have data on the amount of a particular food consumed at one time. In addition, within any population subgroup, a distribution of consumption will exist for any particular food. Some approaches to regulating pesticide residues in food use the 90th percentile of consumption to protect the individuals who eat more food than the average for the population. For chemicals that produce toxicity after multiple exposures, data on average food consumption over time would be more relevant.

Pesticide residues occur in food as a result of commercial agricultural practices and, in some cases, as a result of home gardening. Commercial agricultural practices are generally regulated by national governments, and are influenced by international groups such as the WHO/FAO Joint Meeting on Pesticide Residues (JMPR). These groups generally recommend Acceptable Daily Intakes (ADIs) and Maximum Residue Limits (MRLs), which are the upper limits of pesticide residue expected on raw agricultural commodities consistent with good agricultural practice. In general, these MRLs are mainly for enforcement purposes, that is, to assure that pesticides are not used in excessive amounts. However, since the MRL (a tolerance) is an upper limit, it is not a useful value for calculating actual dietary exposures. In the US, this value is legally defined as a 'tolerance' and is established by the Environmental Protection Agency (EPA).

An accurate estimate of exposure should reflect the effects of washing, peeling, processing, and cooking on pesticide residues. In general, these steps will produce a large decrease in the amount of pesticide present in food at the time of consumption, compared to the amount of residue present in the field. In certain other cases, a pesticide may actually concentrate during processing, or it may be converted to toxic metabolites. The ethylene *bis*dithio-carbamate (EBDC) class of fungicides is known to form the toxicant ethylene thiourea (ETU) during the processing of food treated with these fungicides. Similarly, the growth regulatory daminozide will form unsymmetrical dimethyl hydrazine (UDMH) upon processing (e.g., conversion of apples into apple sauce). In these cases, sampling of field residues alone would not provide an accurate estimate of exposure at the time of consumption. Rather, measurements of field residues combined with studies on the effect of processing and cooking are necessary to accurately predict final residue concentrations at the time of consumption. The residue estimate should also reflect the extent of use of the pesticide. An MRL may be established for a chemical on tomatoes; however, every tomato grown and marketed will not be treated with that chemical.

In addition to residues on fresh and processed produce, pesticide residues may occur in meat, milk, and eggs as a result of feeding items treated with pesticides to domestic cattle and poultry. In certain cases, pesticide residues can occur in meat and milk at significant concentrations. In one instance, dairy cattle were inadvertently fed grain that had been treated with the insecticide heptachlor, resulting in the appearance of heptachlor in milk. In another instance, pineapples treated with heptachlor produced no appreciable residues in the edible fruit; however, when byproducts from a pineapple cannery were fed to cattle, heptachlor residues were detected in milk. In many countries, wastes from food processing are routinely used as cattle feed, resulting in a significant potential for producing secondary residues in domestic food animals.

This issue is especially important where residues occur in milk, since milk products form a significant part of the diet for infants and children, who will also receive an exaggerated dose because of their higher rate of food consumption relative to body weight. In many countries, food has been sampled at the consumer level and analysed for pesticide residues as a means of estimating dietary pesticide exposures. Such techniques include 'market basket surveys', which monitor residues in food purchased from stores and markets, and 'dual diet studies' in which a portion of prepared food is analysed prior to consumption. These approaches provide the most accurate data on pesticide residues at the time of consumption, but are limited by the cost of obtaining and analysing large numbers of samples necessary to achieve statistical validity. For most individuals, pesticide residues in the diet occur as a result of agricultural practices. However, some individuals may receive significant dietary exposures from other sources. The organochlorine pesticides, which are quite persistent in the environment, will concentrate in fish because of bioaccumulation. Commercial and sport fishermen, who regularly consume such fish from lakes or rivers that have been contaminated with these pesticides, may receive a significant dose of the chemical. Hunters who consume wildfowl may receive similar exposures. In one instance, wild geese were discovered to contain significant residues of the organochlorine endrin as a result of eating grain from fields treated with this pesticide. Some birds died from this exposure; surviving birds that were later eaten caused human illness.

The overall impact of such episodes is difficult to assess. Because few individuals within a population obtain their food from these 'wild' sources, statistical population surveys of food consumption will not provide reliable estimates of consumption. However, fish and wild game can be a source of significant hazard to the individual, and should not be overlooked in an assessment of the impact of environmental contamination.

Home gardening, although not a major source of food in most households, may also present a significant hazard to an individual. Generally, the same laws and regulations apply equally to home and commercial agricultural practices. However, some home gardeners, because of a lack of knowledge or through carelessness, have occasionally introduced significant pesticide residues into a food, resulting in illness. Finally, drinking water is a source of dietary pesticide exposure that has become of concern in many nations. In agricultural areas of high pesticide use, sampling of ground-and surface-water supplies has generally revealed the presence of pesticides. The problem may become especially serious in areas with sandy soil, which is easily penetrated by pesticides, and/or areas with relatively high water tables. For example, in Suffolk County, aldicarb was found in about 25 wells sampled at concentrations of up to 515 p.p.b. In California, a state with pesticide manufacturing facilities and a large number of agricultural operations, more than 50 pesticides were detected in the ground water of 23 counties. One pesticide, DBCP, was found in over 2000 wells (25 per cent of the total number sampled), with the highest concentration of 1.24 p.p.b. near a DBCP manufacturing facility. Efforts are currently under way in many nations to monitor ground-and surface-water for pesticides and other chemicals. The monitoring of water for pesticides must consider the largely seasonal application of pesticides. For example, agricultural runoff may produce relatively higher concentrations of herbicides early in the growing season, since pesticides are frequently applied to bare ground, which has a greater potential for runoff. The utility of such monitoring efforts is clear to assess hazards to a population resulting from dietary pesticide exposure. For some subgroups with high water consumption, such as infants fed formula prepared with water, such data are vital to an accurate assessment of dietary exposure.

Ecosystem Monitoring

Of the numerous issues confronted in the monitoring of something as diverse, variable, and complex as an ecosystem, the most difficult is sampling. Adequate principles and guidelines exist for sampling human environments for epidemiological studies; nothing comparable exists for non-humans.

Because of the concern for incidental dietary intake by humans of pesticides, considerable attention has been given to developing accepted procedures for sampling foodstuffs for chemical analysis. Guidelines, indicating substantial agreement over procedures to sample foods, exist in international commerce. The Codex Committee on Pesticide Residues has been prominent in this quest for international agreement and standardization. No such agreement has been reached with regard to ecosystem sampling. The OECD program on wildlife sampling and monitoring partially dealt with the matter, but no guidelines or procedures emerged from this study. Environmental sampling has also been the subject of an IUPAC study.

The paucity of guidelines on ecosystem monitoring, the natural complexity and variability of ecosystems, and the diversity of monitoring methodology makes comparisons of result among investigators virtually impossible. Many existing environmental sampling procedures are carryovers from the randomized, unbiased sampling procedures employed in agriculture to measure endpoints such as crop yields; standard textbooks are devoted largely to such sampling techniques. Environmental sampling requirements are similar to those of epidemiology, namely, reliance on biased non-random sampling, which applies to both spacial distribution and time. For instance, the impact of aerial spraying of a forest on NTOs is likely to be maximal at treetop height and just after spraying; similar considerations apply to the use of pesticides to control biting flies in a fast-flowing river. The failure of some laboratory model systems to extrapolate results to the field may be largely attributed to the failure to include sampling criteria.

The basic need for guidelines for environmental sampling would best be met by a workshop organized by an international agency such as UNEP or OECD. The membership of such a workshop should include a broad array of expertise with particular representation among those specialists who make the greatest use of monitoring data.

Chapter 3

Fungal and Bacterial Leaf Diseases

Symptoms: Leaf spot, tar spot, leaf blotch, leaf blister and shot hole are symptoms of leaf diseases. Shot hole occurs when dead leaf tissue falls out, leaving a hole. *Cause*: A number of leaf diseases that occasionally damage ornamental plants are caused by fungi and bacteria. Many species of fungi cause leaf diseases. Leaf diseases cause leaves to fall prematurely, leaving trees and shrubs without the green factories that produce food for them. Repeated defoliation of trees or shrubs makes them progressively weaker. Most deciduous plants can withstand several defoliations without serious damage, but broad-leaved and narrow-leaved evergreens usually do not recover from a defoliation. Most fungi and bacteria that cause leaf diseases require a wet leaf surface for an extended time, usually about 24 hours. The wet leaf surface allows the fungal spores to swell, germinate and penetrate the plant and the bacteria to swim to a natural opening in the leaf surface, such as a stomate.

Leaf-infecting diseases are more severe if the fungus or bacterium is present at bud break, when the leaves are tender and new. If the weather is dry during bud break, infection occurs later in wet weather after leaves are expanded. Late infections may be unsightly but seldom harm the plant.

Bacterial Leaf Spots

Symptoms: Bacterial leaf spots may appear similar to fungal leaf spots, so it may be difficult to distinguish the two. Leaf spots caused by bacteria are often initially light green and look "water-soaked". Later, these leaf spots usually turn brown or black and may have definite margins.

Cause: Many species of bacteria may cause bacterial leaf spots of landscape plants. These bacteria are often splashed from the soil onto wet foliage, where they enter a leaf through its breathing holes (stomates) or wounds. Thereafter they spread from leaf to leaf when plants are watered or during rainy periods. Bacterial leaf spot of English Ivy is a good example of a common bacterial leaf spot.

Control: Avoid using high nitrogen fertilizers on susceptible cultivars. Only fungicides containing copper may be effective against bacterial blights and bacterial disease, but copper may burn foliage in some cultivars. The antibiotic streptomycin is used in commercial agriculture, but is not available to the homeowner.

Fungal Leaf Spots

Symptoms: Leaf spots are characterized by brown or black spots randomly scattered across the leaf. The spots may or may not have the appearance of concentric rings. The margins of the spot may be a different colour than the centre of the spot.

Cause: Leaf spots are caused by many different fungi that attack landscape ornamentals. Rainy periods favour their sporulation and spread to unaffected foliage. Infection may occur at any time, but the disease is usually most severe on new, emerging foliage during rainy periods.

Control: Fungi that cause leaf spots form spores on the leaves. The new spring infections are commonly caused by spores from last fall's leaves. Therefore, rake and burn or compost last season's leaves before buds break in the spring. Help foliage dry quickly after rain, heavy dew, or sprinkler watering. In the case of deciduous trees and shrubs, prune out overthick growth for better penetration of solar radiation and better air circulation to aid in drying foliage. In the case of conifers, it is most important to control weeds that surround the lowest branches. Weeds inhibit air circulation and increase humidity, favouring fungal infection of needles.

Fungal Sinusitis Treatment: a Life-long Commitment

Types of Disease

Fungal sinusitis can be invasive or noninvasive, said Richard D. deShazo, MD, director of the Division of Allergy and

Immunology at the University of Mississippi Medical Centre in Jackson. In noninvasive disease, fungi appear only in mucus from the sinuses. When the disease is invasive, fungi may penetrate the sinus mucosa, submucosa, blood vessels, and/or bone. Noninvasive fungal sinusitis is most common in patients with allergic rhinitis, asthma, chronic sinusitis, and nasal polyps. These patients, many of whom have undergone multiple sinus surgeries before diagnosis, may have AFS or sinus mycetoma. Diagnosis hinges on histopathologic visualization of fungal elements. For instance, allergic mucin—an eosinophilic, fungi-containing, peanut butter-like substance—develops in the sinuses of patients with AFS. Although other forms of fungal sinusitis can have a thick or gravel-like exudate, only AFS manifests the characteristic allergic mucin.

Sinus mycetoma may occur in any paranasal sinus and, unlike AFS, is often unilateral. Sinus mycetoma patients may present with symptoms of chronic sinusitis. "Some intermittently blow gravel-like material from their noses," added Dr. deShazo.

Invasive fungal sinusitis, which primarily affects immunocompromised patients, exists in three forms: granulomatous invasive fungal sinusitis, chronic invasive fungal sinusitis, and acute fulminate invasive fungal sinusitis. Granulomatous invasive fungal sinusitis occurs almost exclusively in North Africa; it may produce pressure necrosis of the sinus wall, orbital proptosis, or even brain erosion. Chronic invasive fungal sinusitis is characterized by a low-grade, mixed cellular infiltrate; nasal polyposis; and thick, purulent mucus. Decreased vision, impaired ocular mobility, and proptosis may develop if the infection reaches the orbital apex. Acute fulminate invasive fungal sinusitis is caused by rapid spread of fungi from the nasal mucosa and sinuses into the orbit, vessels, and brain parenchyma. "Prognosis is poor," said Dr. deShazo, "although with aggressive and early treatment, up to 50% of patients may survive."

All Fungi are Pathogenic

Patients are increasingly becoming "living Petri dishes," said Michael G. Rinaldi, PhD, a clinical mycologist. The increasing prevalence of immunocompromised patients—the result of human immunodeficiency virus infection and of treatments that impair the immune system—has meant a growing incidence of mycotic infections, including fungal sinus infections. The list of fungi that can infect the sinuses is long. "Basically, there are no longer any nonpathogenic fungi," indicated Dr. Rinaldi, director of the Fungus Testing Laboratory and a professor of pathology, medicine, and microbiology at the University of Texas Health Science Centre at San Antonio. "These days, given the right immunocompromised person, virtually any fungus can cause human disease." The unique case of a 26-year-old man with end-stage human immunodeficiency virus infection and a low CD4 T-cell count provides an illustration. The patient, Dr. Rinaldi recounted, unknowingly inhaled spores from the wild mushrooms he had gathered and prepared for a meal. Several weeks later, the patient presented, complaining of itchiness and a fissure in the roof of his mouth. His physician discovered the head of a mushroom literally growing inside the patient's head. The mushroom protruded down through the fissure from the sinus cavity. Culture confirmed it was the same type of fungus—*Schizophyllum commune*—that the patient had prepared for the meal.

Aspergillus fumigatus and *A flavus* are much more common causes of fungal sinus infection, and they may be lethal. Under the microscope, *A fumigatus* has a smoky, gray-green appearance, said Dr. Rinaldi. In contrast, *A flavus* is lime green in colour. In addition to *Aspergillus* sp, *Bipolaris spicifera* and *Curvularia lunata* often cause AFS.

Identifying some of these organisms is difficult because they closely resemble numerous other types of fungi, including some molds that may be equally deadly. Physicians should therefore always use a good microbiologist to identify the infecting organism when fungal sinusitis is suspected, Dr. Rinaldi recommended, not only because it is good medicine, but because such identification may help determine the most appropriate therapeutic approach.

High Recurrence Rate

A variety of surgical procedures exist for the treatment of fungal sinusitis. The presence of acute fulminant invasive fungal sinusitis constitutes a surgical emergency. "It requires immediate, radical resection of tissue and application of systemic antifungal agents," emphasized Dr. Kuhn. Without such intervention, the patient has a poor chance of survival. The previous treatment for chronic sinusitis was open sinus surgery, such as frontal sinus obliteration or external frontoethmoidectomy. A major disadvantage of open surgery, however, is that it typically involves removal of large amounts of bone and healthy sinus tissue. The newer intranasal sinus procedures, such as functional endoscopic sinus surgery, are usually far less damaging. "The purpose of functional endoscopic sinus surgery is to remove the obstruction at the outflow tract of the sinuses while preserving all possible mucous membranes and all possible normal structures," according to Dr. Kuhn.

Regardless of the procedure, recurrent disease is common—usually, as a result of incomplete surgery that leaves obstructions likely to impair mucocilliary clearance. For example, the recurrence rate of chronic sinusitis is 55% when a poorly resected uncinate process obstructs the maxillary sinus ostium; it jumps to 80% when residual infraorbital ethmoid cells produce such an obstruction. Postoperative disease recurrence is also highly likely with allergic fungal sinusitis. In a study of 24 patients with the disease, eight of the nine who underwent surgery had a recurrence; the ninth patient experienced a recurrence after the study was completed. "So the key take-home message," said Dr. Kuhn, is that "every patient who is operated on and gets no medical treatment afterward will suffer a recurrence. You just have to wait long enough."

The two longest periods between surgery and recurrence were 29 and 34 months, suggesting that fungal sinusitis patients require very long-term follow-up. In practical terms, the extreme likelihood of recurrence means "that I need to follow these patients for at least five years," Dr. Kuhn concluded. He recommended monthly follow-up for all patients during the first year after surgery, then bimonthly follow-up during subsequent years for those who are doing well.

Corticosteriods Reduce Recurrences

Further study is required to determine the optimal pharmacotherapy for fungal sinusitis. However, starting long-term oral corticosteroid therapy immediately after surgery appears to reduce recurrences and disease severity, at least with AFS, said Dr. Schubert. By way of evidence, he described a retrospective study of 67 AFS patients who received standard postoperative treatments (eg, anti-inflammatory nasal sprays, antihistamines, decongestants, and allergen immunotherapy) or standard treatment plus oral corticosteroids.

The corticosteroid group received prednisone; 0.5 mg/kg/d was given for two weeks after surgery, followed by 0.5 mg/kg every other day for about three weeks. The dosage was gradually tapered to between 5 and 7.5 mg every other day by three months, and then 5 mg every other day was given for the remainder of treatment (typically, about a year). If repeat surgery was necessary, this protocol was restarted from the beginning. At two months' follow-up, most patients in the corticosteroid group had mild or no AFS, whereas patients in the standard treatment group typically had moderate or severe disease, Dr. Schubert said. The results were even better at one year. The rate of repeat surgery for recurrent disease at 200 days' follow-up also reflected the advantage of corticosteroid use: It was less than 20% in the prednisone group versus about 50% in the standard treatment group. Oral corticosteroids provided significant benefit, stressed Dr. Schubert, even when patients did not complete the entire treatment course. Serial measurement of total serum immunoglobulin E (IgE) levels has prognostic value, the study also showed. Patients with fungal sinusitis typically exhibit high-titer total and fungal-specific serum IgE levels. "If that level increased 10% or more, the chance of needing recurrent sinus surgery was roughly 50%," Dr. Schubert reported. Furthermore, the total serum IgE level corresponded with the severity of AFS, rising when the disease worsened and falling when it improved.

Dr. Schubert also emphasized the importance of using a "steroid burst" (a temporary increase in the prednisone dose, followed by a rapid taper to its normal level) whenever a patient with AFS has an intercurrent acute upper respiratory tract infection. "Absent that maneuver," he said, "patients with acute upper respiratory tract infections have an increased tendency to relapse." Systemic antifungal drugs have not been found beneficial, possibly because AFS is noninvasive, Dr. Schubert suggested. It is unknown whether the disease responds to topical antifungals, which are problematic anyway because of the difficulty of applying them to sinus tissue.

Organic means of Disease & Insect Control

Dormant Spraying

Dormant spraying means exactly what it says; spraying done when the plant are in dormancy. That means no leaves or active growth should be happening. Dormant spraying should be done immediately after pruning in spring and again in late fall when the leaves drop (usually late Oct.) Spraying dormant oil & lime sulfur when there is active growth will burn & damage leaves so if you snooze you loose in this instance! Your dormant spray will damage that tender new growth, setting your plant back, and possibly resulting in deformed growth.

A thorough dormant-spraying offers you your best chance for control of the various fungal diseases that plague roses: powdery mildew, rust, blackspot, insect pests, even downy mildew-all can be reduced by a careful dormant spraying.

Dormant Oil sprays and Lime-sulfur sprays are available at our nursery, and they are very reasonably priced. These materials are usually formulated for combined use. They may be safely used on roses now, while there is no foliage to be damaged. These oils coat the canes & suffocate the eggs & adult insects so spray each rose plant thoroughly, and spray the ground around it as well. Totally saturate the canes making sure you have not missed a minute crevice! Insects & tiny dust like spores of diseases hide in the tinniest of crevices & will come to life & multiply if you miss them. This application is essential if you are having spidermite problems. Those little unbearable pests are very hard to get rid of & a double dose of dormant oil really works wonders. In the fall & again in the spring giving them a double whammy.

Insecticidal Soap

When using insecticidal soaps, spray the entire plant thoroughly and repeat applications frequently. You may have to spray every few days at first until the infestation is under control. Apply in the morning, evening or on overcast days as not to burn tender new leaves if sprayed in full sun. Read carefully all label instructions. Soaps can burn some skin sensitive people or hurt eyes. If mixed too strong can cause severe damage to tender new growth! Rugosa Rose leaves do not like to be sprayed with anything especially insecticidal soaps! Spray only the parts where insects are a real problem. If possible use rain or soft water as soaps adheres better in water without high minerals content.

Organic Tips

- To prevent spreading fungal diseases from plant to plant, disinfect pruning shears between cuts, dip the blades in a 10% solution of chlorine bleach(1 part bleach to 9 parts water) or pure alcohol or peroxide Some rose growers submerge their bareroots in the bleach solution or peroxide before planting in spring or fall. Only dip them do not leave them in it for any length of time. This works really quite well! Canes need to be completely coated so dipping is the best! Pheromone traps use specific (sex hormones) attractants to lure insects inside where they get stuck or trapped. They are also used to monitor and alert the grower when insects appear on the scene. Sticky cards are a great help for insects such as whitefly (yellow) and midges (blue cards).
- As unfortunate as it is, sometimes we have to eliminate the source of our problems! Getting rid of the troublemakers although painful, is one way to help rid and curtail diseases; especially if you want to grow organically! Removing the varieties that are susceptible (prone) to diseases will help remove the source of infections and overall disease problems. This happened to us early in our nursery business and changed our outlook on what we grew there after! As we loved and grew some "Soel D'or" and "Persian Yellow" hybrid Rosa foetidas very prone to blackspot and more so in wet Nova Scotia climates! After continual sprayings of sulfur we came to the conclusion of roses that are really prone to disease should not be grown period! Especially if your a nursery that has to keep all its stock in top shape. You just can't win with varieties like these, as beautiful as they may be. Be sure to use dormant oil

& sulfur the fall & spring after to eliminate all residues left in beds & other infected roses.

- Spraying with broad-spectrum insecticides (those that are effective on ALL insects) reduce many natural predators that feed on the problem pests also. When you use these products you're wiping out pretty much ALL INSECTS where you have sprayed! This includes MR. Bee to the left! Population increases have been studied & noted on the occurrence of spider mites & thrips and the use of broad-spectrum insecticides on them. Reason, Not only are the natural Enemies of these pests killed, but also newer generations have built up a greater resistance to the insecticide and future applications will be less & less effective.
- A strong jet of water from a garden hose can help control many diseases and insect pests. A direct stream to the tips where aphids are and the undersides of leaves where spidermites flourish will knock and keep these critter at bay. Aphids are very slow moving insects and the chances of those ones getting back on the plant is almost nil due to their predators on the ground and spidermites absolutely HATE WATER, so they won't hang around where they are constantly subjected to this. Because wet foliage encourages fungal diseases, it is wise to do this early in the day or on a windy day so leaves dry off readily. Never at night, if you can help it.
- Blackspot, mildew & insect problems can be reduced by cleaning up and destroying all infected leaves early in the infestation, (Do NOT compost them) Good fall housekeeping is essential to not carry it on into next season. Clean-up all fallen leaves and leaves still on roses. Again don't compost these burn them if you can.
- Botanical insecticides include rotenone and pyrethrum. Although these insecticides are derived from plants, they are not necessarily safe. In a well-managed garden, these should be used only when other methods have failed. They kill both beneficial as well as pest insects. They should be sprayed only at dusk when fewer

honeybees are active; rotenone and pyrethrum are toxic to fish and should not be applied near ponds. Rotenone is an insect stomach poison best applied as a 1% solution in water. Pyrethrum is a nerve poison and paralyzes insects on contact. Nicotine, another botanical insecticide, is best not use it is very highly toxic to mammals.

- Organic sprays and dusts include insecticidal soap and diatomaceous earth. Both are effective against soft-bodied insects such as aphids, spider mites, and some caterpillars. Insecticidal soaps penetrate insect membranes, causing paralysis and starvation. Use soft water when mixing and add isopropyl alcohol (2 Tbsp. per liter). Diatomaceous earth is a dust composed of the fossilized skeletons of microscopic sea organisms. The individual particles are sharp and cut insect bodies, causing them to dehydrate and die. It is applied as a dust, or in a water solution. Homemade sprays of ground up garlic and hot peppers can also effectively repel some insect pests.

Cultural and mechanical practices involve handpicking insect pests when they are few in number and crushing them underfoot, keeping the yard and garden free of decaying refuse, rotating the position of crops each year, and maintaining healthy plants with fertile soil.

Bacillus Thuringiensis, "BT"

1. Bt (Bacillus thuringiensis) The Natural Insecticide Bacillus thuringiensis (Bt) is a naturally occurring, soil borne organism that is used for its ability to control certain insect pests especially larvae of flies and months that can severely damage roses and other plants. Bt works in a natural, environmentally friendly manner. Bacterial agents, like Bt, are effective in controlling insects in the larva stage only. In this stage the larvae is usually a caterpillar or worm and to effectively control the pest, you must get it early at this stage both to eliminate the pest and to avoid too much damage to the leaves of the rose. The Bt is applied to the foliage of plants infested with a leaf or needle-eating larva. It is

not as important to completely saturate the leaf surface as with insecticidal soaps as the larva need only eat a portion of the leaf with the bacteria on it to do the trick. You'll usually find these critters on the undersides of leaves and usually the same colour as the leaf as not to attract their predators therefore Bt should be applied to the UNDERSIDES of leaves.

2. Bt is also broken down faster in sunlight. Application to the undersides of the leaf surfaces will prolong Bt's activity. High temperatures do not encourage Bt's breakdown.
3. After Bt spores are ingested by larvae, they grow and reproduce, meanwhile producing crystalline toxins. The crystalline toxins paralyze the digestive tract of the larvae causing it to cease eating. Death will follow, but time of death can range anywhere from 12 hours to 5 days after ingestion This depends on the amount of Bt ingested, the size and variety of the larvae and variety of Bt used for control. There are different strains or varieties of Bt available that have been selected for the control of specific insects. The one we promote for roses is Bt variety kurstaki (BTK). This controls the European corn borer, tomato hornworms, fruitworms, cabbageworm, cabbagelooper, spring and fall cankerworm, spruce budworm, and other caterpillar-like larvae. Other Bt varieties like San Diego (BTSD) controls early larvae of the Colorado potato beetle and Bt variety israelensis (BTI) controls mosquitoes, black flies and fungus gnats.

Bt was used extensively in Nova Scotia's forests these years for control of the Tussock Month as great treat to the evergreen forest and Christmas Tree Industry. The Bt has gained well-earned popularity because of its distinct advantages over other pesticides such as :

(i) Hazards to humans are negligible although inhalation or contact with eyes or open wounds should be avoided.

(ii) Using it on vegetables, Bt can be used right up until harvest, which allows for a longer-term control as compared to other insecticides requiring a waiting period from time of application to time of harvest.

(ii) There is no waiting period from time of application before re-entering the field.

(iii) Ficial or non-target insects are not harmed.

(iv) Insects that ingest the Bt and later die from it are not dangerous to birds or other animals that may feed the dead insect.

(v) Is not known to cause injury to plants on which it has been applied and is not considered harmful to the environment.

(vi) With any substance you should not depend and use only it as insects tend to build up resistancy over time to them and for the most effective use try to use other means also every now and then.

Organic Earwig Control

Earwigs can really be a menace and to know how to effectively control them organically you must first know their life cycle. Here in Canada the earwig over winters as an adult in nests that is made in cracks and crevice, under rock and boards around the yard.

The female lays a portion of her eggs in the fall. In the spring she will finish laying her eggs in the nests she has created. In late March to April, depending on how warm it gets, the males leave the nest. This is why early in the year one will see a few large earwigs. The female will remain in the nest and care for the eggs and the hatched young until late June to July. At this time the young earwigs and the females will leave the nests looking for food. This is the time of year when most people find that earwigs are a problem. This then is the time to focus on controlling them.

Try to confirm the life cycle with your area Dept. of Agriculture knows about when the females open up the nests. The most effective way to control earwigs without insecticides is the use of trapping.

Again knowing the insects habit makes the use of traps fairly effective. Earwigs are night creatures. They feed and move about in the dark. During the daytime they will find areas to hide from the light. Any small crevice or container makes great daytime homes. Traps should be set in the evenings and collected in the morning to remove the captured insects. The best types of traps that seem to work well are: corrugated cardboard; Rolled up magazines or newspapers; small cans with openings punctured in the ends.

Make sure holes are large enough to let the earwig in tubing such as bamboo rods. Two pieces of wood with groves cut and taped together. When collecting the traps, make sure they are collected each day in the mornings. Dispose of the earwigs in a container of diesel fuel or gasoline. This will kill the insects quickly.

When placing traps, make sure they are placed in and under shrubs and other dark hiding places that the earwigs would like to stay. Place the traps when the nests are opened up later in the summer. People that report failures with traps result from setting the traps too early. They operate the traps and find that no or very few earwigs are caught. They then stop trapping and when June/July comes earwigs by the "hundreds" are found and they think the traps were not successful. "Wrong" just poor timing and not working along with the earwigs life cycle.

Fungal Leaf Spots

Brown spot is a common disease which occurs frequently on young plants. If warm moist weather conditions continue, infection, and defoliation progress from lower to upper leaves. Usually the hot dry weather of midseason arrests disease development, but the disease may become active again near maturity. If conditions favourable for disease development continue into the season or reoccur before maturity, the resulting defoliation can cause serious yield reductions. The pathogen overwinters on infected leaf and stem debris, and the disease is most severe when soybeans are grown continuously in the same field. Cotyledons, primary leaves, and lower trifoliolate leaves typically show brown spot symptoms. The brown to red lesions vary in size from pinpoint to 4mm wide, but may coalesce and form larger irregular shaped spots. Severely affected leaves turn yellow and drop. Although there are differences in susceptibility of cultivars, no highly resistant cultivars are available. Control is best achieved by rotation and deep plowing to reduce disease incidence.

Downy mildew, one of the most common foliar diseases of soybean, occurs worldwide, but is seldom a serious problem. This disease reduces seed quality and seed size, if extensive defoliation occurs, yields can also be reduced. Disease development is favoured by high humidity and cool temperatures.

Symptoms are found on young plants, but the disease does not become widespread in a field until the late vegetative or early reproductive stages. The increased resistance of older leaves and higher temperatures at mid-season stops disease development before extensive damage occurs.

Lesions start as small light green to yellow indefinite spots on the upper leaf surface. As spots enlarge they become yellow-brown to dark brown. A grayish downy tuft of sporangiophores develops on the lower leaf surface during periods of moist weather. Severely infected leaves drop prematurely. Cultivars with some resistance are available and should be used. However, many races of the pathogen have been identified, and cultivars which are resistant to all known races have not yet been developed.

Powdery mildew occurs sporadically on soybeans. Yield reductions are possible if foliage is affected during pod fill or extensive defoliation occurs. Plants are susceptible at any growth stage, but symptoms are rarely seen in the field until the mid to late reproductive stages.

Disease incidence is greater in seasons with cooler than normal temperatures. The first symptom is usually a circular area of white to light-gray, superficial powdery growth on the upper leaf surface. Under favourable conditions, cool humid nights and mild dry days, these areas expand rapidly, coalesce and cover the entire leaf surface. All green plant parts may be covered by the powdery growth. Heavily infected leaves turn yellow, then brown, and drop prematurely.

Powdery mildew is controlled with resistant cultivars. Most determinate soybean cultivars are highly resistant to powdery mildew, while many indeterminate cultivars are susceptible.

Aerial blight is found in several states, but is primarily of economic importance in Louisiana where extensive yield losses have been reported. The fungus survives in the soil and on plant debris, as well as on certain weed hosts. Infection usually begins at flowering during prolonged periods of high humidity and warm temperatures. Symptoms include leaf spots, leaf blighting, or defoliation. Leaf spots are irregular water-soaked spots that develop reddish-brown margins during drier conditions. Tissue in the centre of lesions falls out giving the plant a ragged appearance. Brown lesions may also form on the petioles, stems, and pods. Severe infection causes pod blight.

Control strategies include the use of good management practices (seed bed preparation and weed control) and applying foliar fungicides. Some cultivars are more tolerant and should be planted in fields with a history of aerial blight. This destructive disease occurs regularly in areas with high humidity. In other areas, the occurrence of frogeye is sporadic, but is more severe in seasons with frequent rainfall. When a high percentage of leaves are heavily infected, serious yield loss is common. The use of resistant cultivars has reduced losses.

Young expanding leaves are extremely susceptible, while fully expanded leaves are more resistant to invasion. However, because visible lesions develop nearly two weeks after invasion, lesions are never seen on young expanding leaves. Leaves that expand during periods of weather unfavourable for invasion remain relatively disease free and result in "layers" of healthy leaves between "layers" of heavily diseased leaves. The foliar lesions measure up to 5mm in diameter but may coalesce to form larger spots. Young, fully developed lesions have a gray to brown centre with 'a distinct purplish to reddish-brown margin. As lesions age, the centre becomes tan to nearly white and the margin darkens. Lesions also occur on stems and pods. Crop rotation and plowing under crop residues will reduce disease incidence. Fungicides applied at the late flowering and beginning seed growth stages will protect against frogeye infection. Several races of the pathogen have been identified, but cultivars with resistance to all known races are available. Cercospora leaf blight is a late season disease that is often mistaken for early senescence. Symptoms are first observed at the beginning of seed set. Extensive losses from the disease have been reported in some years. While this fungus also attacks pods and causes purple seed stain, there is no consistent relationship between the occurrence of leaf spot and purple seed stain. Symptoms are first observed at the beginning of seed set. Small, reddish purple, angular to irregular lesions occur on the upper leaves which become leathery and dark purplish-red with bronze highlights. This discoloration may extend to the upper stems, petioles, and pods. Heavily infected leaves rapidly turn yellow and fall, mimicking senescence. Lower leaves remain green.

Although some cultivars possess resistance, control is primarily achieved by application of foliar fungicides starting at early pod set. Alternaria leaf spot has been reported from all soybean growing areas. This fungus is often a secondary invader following mechanical injury, insect damage, or another disease. Altemaria leaf spot is a late season disease that develops on maturing leaves or on younger leaves if injured. If uninjured, most improved determinate cultivars appear to be resistant until leaves start to senesce. Since lesions develop after the onset of yellowing, yield losses are considered minimal. The causal agent appears to be seedborne.

Lesions are nearly circular unless they are restricted by a major vein or coalesce with another lesion. Single lesions range up to 25mm in diameter, have brown concentric rings, and a well defined margin. Control measures are not considered necessary. Target spot occurs regularly in low incidence in many areas.

The disease usually occurs too late in the season to cause economic damage. However, the disease is potentially serious on late-maturing cultivars or on susceptible cultivars in seasons of high rainfall. Leaf symptoms are seen prior to maturity during periods of high humidity or when free moisture is present.

Lesions on leaves are reddish-brown, circular, and variable in size up to 15mm in diameter. Spots may also be found on petioles, stems, and pods. The larger lesions often show distinct concentric zones of dead tissue and may have a narrow, indistinct yellow-green halo.

Lesions which form along veins may be narrow and elongated. Severe infections may cause premature leaf drop.Most well adapted, high yielding, determinate cultivars have tolerance to target spot.

Chapter 4

Virus Diseases of Forage Feed Plants

Couch Grass-Uses and Benefits

Herb Description

Couch Grass (Elymus repens) is a very common species of grass inherent to Europe.It has found throughout much of the world. It is a recognized as a notoriously It has creeping rootstock which enable it to grow rapidly across grassland. In summer time evergreen couch gives a beautiful green surface layer of ground containing a matt of grass and grass roots.Couch grass has hair like leaves & straight up flower spikes, Its long, yellowish white rootstock must be completely dug up to eradicate the plant because broken rootstock produce new plants.

The flower spike is 4 to 7 inches long, with spike lets almost one inches long, 5 to 7 mm broad & 3 mm thick. The stalk of plant mature to 47 inches tall, with linear leaves 6 to 16 inches long and more then half inches broad at the base.

It flowers in June to September. The flowers are hermaphroditic and are pollinated by Wind. The plant likes sandy, loamy and heavy soils for growth. The plant choose acid, neutral and alkaline soils and can develop in very acid and very alkaline soils. It requires dry moist or wet soil.

Properties

- Soothing mild diuretic.
- Demulcent.
- Antimicrobial.

- Aperient.
- Anticholesterolaemic.

Uses and Benefits

Couch Grass is a useful remedy in the treatment of various problems such as bronchitis, colds, cough, fever, sore throat, tendency to infection, urinary tract infections.

Traditional peasants in South China and Hong Kong are still use Couch Grass roots by make a herbal tonic tea. Couch grass is used externally & applied as a wash to swollen limbs

In Europe grass roots are used to make a popular drink to rid of impurities the blood. It had many uses in traditional medicine, especially for kidney ailments. The grass roots are aperient, antiphlogistic, demulcent, diuretic, emollient, lithontripic and tonic actions,is used treat some cases of urinary incompetence & as a worm expellent.

Couch grass is an advantage as a herbal medicine to treat wide range of kidney, liver and urinary disorders

Couch Grass is a valuable remedy in the treatment of various disorders such as:

- Bladder infection.
- Bright's disease.
- Bronchitis.
- Catarrh.
- Constipation.
- Cystitis.
- Diuretic.
- Emollient.
- Fevers.
- Gallstones.
- Gout.
- Jaundice.
- Back pain.
- Rheumatism.
- Skin diseases.
- Stones.

- Sudorific.
- Syphilis.

Side Effects

Couch grass is generally thought to be a safe herb, with some reported serious adverse events. Few cases of side effects are enhance heart rate & rhythm, nausea, & vomiting have been reported.

There are some side effects may also include soporific & mental slowing. It enhance the risk of bleeding and alter blood tests that measure blood clotting.

Dosage

- Dried root:

4 to 8 gm or in decoction. Take three times in a day.

- Liquid extract:

1:1 in 25% alcohol, 4 to 8 ml. Take three times in a day.

- Tincture:

1:5 in 40% alcohol, 5 to 15 ml. Take three times in a day.

Drug Interactions

Some Scientific studies shows, use of couch grass with alcohol or other soporific drugs may heighten the amount of drowsiness caused by some drugs are ativan, barbiturates, narcotics, some antidepressants.

Fescue-Diseases

Fescue lawns can withstand diseases if they are properly cared for. This includes fertilizing properly, mowing at the right height and watering when needed. It is better to control diseases by proper maintenance than by using fungicides.

There are four common diseases of fescue lawns:

Pythium Blight usually occurs in late spring when there are hot days, warm nights and plenty of rain. You may be able to see the white fungal growth at the edges of a dead patch of grass. It is most common on lawns which have had too much seed planted in the fall or spring. It is made worse by too much rain or irrigation. The crowded condition of the new seedlings allows the disease to wipe out large patches of lawns in a few days. Fertilizing heavily late in the spring causes lush growth for the fungus to grow on. Controlling pythium blight with fungicides is difficult. It is best controlled by attention to seeding and fertilizer rates.

Brown Patch is most prevalent on fescue fertilized in the summer when night temperatures are above 68 degrees. The dead patches may start small but can grow and join together to make patches more than 3 feet apart. To control it, do not fertilize fescue heavily in summer and if you irrigate, do it in the very late evening. There are lawn fungicides available to control brown patch. Read the label carefully and use the rate and timing that is indicated.

Slime Mold is a startling fungus. It looks like someone has poured oil over a tabletop sized area of the lawn. The fungus grows on the grass blades and will be gray before it turns black. Fortunately, the fungus does not attack the grass. It is ugly but it can simply be washed off with a garden hose. No fungicides are necessary.

Dollar Spot occurs when nights are cool and days are warm. The spots of dead grass are about the size of your hand. It is associated with a lack of fertilizer and drought conditions. To control it, apply a moderate amount of fertilizer and irrigate deeply once per week. There are lawn fungicides available to control dollar spot. Read the label carefully and use the rate and timing that is indicated.

Fungicides If you are absolutely sure you have a disease, the disease can be controlled with fungicides. Chlorothalonil (Daconil) and Bayleton (Fung-Away) are labelled for lawn disease control. Although they are effective, their cost may cause you to reconsider their use. To cure brown patch in a lawn requires an application of fungicide every 14 days. A lawn fungicide costs approximately $20 per 1000 square feet per application. For a typical 5,000 square foot lawn, that comes to $100 every two weeks. Let your pocketbook be your guide!

Lawn Disease ID: If you suspect that you may have a turfgrass disease, your sampling procedure is critical in order to get an accurate diagnosis. Go to the suspected disease area. Find where the live grass meets some dead grass. This is called the disease margin area. You need to cut a 4 X 4 inch square that has half dead grass and half live grass in it. Include the roots. Do not allow the sample to heat, do not allow the sample to dry out. Place in a paper bag or plastic bag which is not sealed. If you cannot bring the sample in the same day you cut it, it is best to refrigerate it. Take it to your local county Extension office. They would prefer to get the sample on Monday or Tuesday so that they can mail it to the plant disease lab without any delay in the mail. Results will be mailed to you in about 10 days.

Fescue Lawn Turf Grass Information

Geographic Zones: Fescue grass grows best in colder climates. For the most part, it will survive well in the majority of central and northern states. Only the southern states, where the average temperatures are high, does fescue have difficulty growing.

Water Requirements: This grass has very low water requirements and should be watered only when the soil is dry or the grass shows signs of needing water. Watering too often will make the grass more prone to diseases and pests.

Fertilizer Requirements: Fertilization requirements vary for each variety of this type of grass, but are all generally low. It is important to keep in mind, however, that for all varieties, it is best to avoid fertilizing much in hotter months to cut down on disease and insect infestation.

When to Plant: While this type of grass can be planted in the fall and spring, early fall is considered to be the best time. In fact, the absolute best time to plant is generally thought to occur between September 15 through November 1.

Recommended Mowing Height: Mowing height is largely determined by the variety of grass planted and the climate of the area. This can range from 1-3 inches in height.

Light Requirements: Just like its water requirements, the light requirements for this grass is fairly low, leading this to be one of the most shade-tolerant of the cool season grasses.

Best Places to Use: This grass does well in low-light, high-traffic situations, which makes it perfect for many different purposes, ranging from personal lawns to playgrounds and sports fields.

Common Diseases: Overall, fescue is fairly tolerant to most diseases that affect grass. It is, however, particularly susceptible to fusarium blight when it is young and, in older lawns, leaf spot and brownpatch can cause difficulties.

Common Pests: These include armyworms, cutworms, and white grubs. White grubs are a particular threat to watch for. If 3 or more are found to exist in the lawn per square foot, then action should be taken to prevent damage as soon as possible.

Soil pH: Recommended pH levels for this grass is from 5.5-7.5

Special Advantages:

- Very durable grass.
- Tolerates low-light and heavy-traffic areas.
- Drought resistant.
- Needs very little maintenance to survive.
- Fairly resistant to most diseases.

Special Problems:

- Does not handle high heat situations well.
- Requires a greater mow height than many other grasses to thrive.

Bacterial Virus Gene Confers Disease Resistance In Tall Fescue Grass

Researchers at North Carolina State University have discovered that inserting a specific gene from a bacterial virus into tall fescue grass makes the grass resistant to two of its biggest enemies

The NC State researchers showed that the inserted gene – the T4 lysozyme gene, a gene found in bacteriophages, or bacterial viruses – conferred high resistance to gray leaf spot disease in six of 13 experimental grasses. Three of the six resistant grasses also showed high resistance to brown patch disease. These two diseases are arguably the most important– and severe – fungal diseases affecting tall fescue grass.

The finding has the potential to have wide applications in engineering resistance to a variety of fungal diseases in not only tall fescue grass – the most widely planted turfgrass in North Carolina and a commonly utilized grass in the southeastern United States – but various other crops. The collaborative research involves four faculty members: Dr. Ron Qu in the Department of Crop Science, Drs. H. David Shew and Lane Tredway from the Department of Plant Pathology, and Dr. Eric Miller, in the Department of Microbiology. The research was mainly performed by Dr. Shujie Dong, a post-doctoral researcher who was a graduate student of Qu's, with assistance from two other scientists in Qu's lab – Drs. Jianli Lu and Elumalai Sivamani.

About half of the turfgrass planted in North Carolina – one million acres – is tall fescue grass, a cool-season grass that has a high tolerance for the heat and drought of North Carolina summers, Tredway says. It is ubiquitous in the Southeast, found on lawns, golf courses and commercial acreages. Gray leaf spot disease is caused by the Magnaporthe grisea fungus, the pathogen that also causes rice blast – the major disease of rice plants. Gray leaf spot causes round or oval tan spots that turn gray when there's high humidity. It infects blades to make the grasses die rapidly.

Brown patch disease, caused by the soil-dwelling fungus Rhizoctonia solani, a major pest to various plant species, brings about circular, brown lesions on grass. Lawns with brown patch disease appear wilted, even if watered sufficiently, the researchers say. Miller, the microbiologist, says that the bacterial viruses exist widely in different environments, and produce an array of products that are harmful to bacteria; as viruses attempt to spread, which they need to do in order to survive and thrive, the T4 lysozyme gene produces the enzymes that chew through the bacterial cell walls. Miller says that the lysozyme now made by the grass does essentially the same thing to a fungus when it tries to infect, thereby providing anti-fungal properties in tall fescue and allowing the grass to withstand fungal disease.

Sorghum Disease Management

Ergot (Claviceps Africana)

Sorghum ergot was first recorded in Australia in April, 1996, and is now endemic in Queensland. Losses in commercial grain crops have been minimal, with tillers flowering early or late in the growing season most affected. Ergot can occur at any time during the growing season. Only a short period of unseasonally cool wet weather during flowering, when pollination and fertilisation are adversely affected, is needed for infection.

Ergot is readily identified by the 'honeydew' oozing from sorghum flowers. This 'honeydew' is not as sticky as that produced by Aphids.

It later dries into a white powder which is often seen on the leaves and on the soil under affected plants. Infection occurs at flowering during cool humid weather when spores land on exposed flowers. The developing seed is replaced by a fungal growth which eventually hardens to a creamy-coloured sclerote (ergot).

Ergot spores compete with pollen for a site on the sorghum flower. Well pollinated crops may have some ergot, but are unlikely to suffer economic damage. Crops with impaired pollination are at risk. Consequently, crops at greatest risk are: forage sorghums, heads on ratoon crops flowering late in the season, tillers on late flowering plants, and crops in which pollination has been affected by a cold snap before or at flowering.

Ergot survives all year round on Johnson grass, *Sorghum almum*, and volunteer grain and forage sorghum. The spores are spread mainly by wind but can also be spread by insects, animals, humans, and machinery. Tillage, stubble burning and crop rotation have little effect on survival.

Ergot can reduce crop yield, cause harvest problems with sticky honeydew, contaminate the grain sample and cause toxicity problems in livestock feeding on heavily contaminated grain.

To minimise the risk of ergot contamination, growers should:

- Sow at recommended times.
- Ensure even flowering (use press wheels, high germination seed, accurate seed depth control, good crop nutrition, wide row spacing (90-100 cm) especially with hybrids that tiller heavily, consider pre-harvest kill-off of tillers using glyphosate herbicide.

- Graze forage sorghum heavily to delay flowering until after grain sorghum has flowered. This reduces inoculum levels and the potential for ergot infection of grain sorghum.

Rust (Puccinia Purpurea)

Symptoms: early symptoms on leaves are small purple red or tan spots. These enlarge to produce elongated raised pustules that break open to release brown, powdery masses of spores.

Sorghum rust is more serious in late sown crops or susceptible hybrids in humid areas. If the disease is serious, leaves are destroyed and pinching of the grain results. Select hybrids with resistance for late planting.

Johnson Grass Mosaic Virus

Symptoms: Mosaic (light and dark green lines on veins), Red Leaf (severe leaf reddening, followed by formation of red spots or large areas of dead tissue, and Red Stripe (red or tan stripes parallel to the veins).

Hybrids developing red leaf or red stripe reaction should not be sown late as the disease can be serious, causing stunting and death in some plants. The virus is spread from plant to plant by aphids. Control by planting resistant hybrids. A strain of the virus exists in south and central Queensland that can infect resistant hybrids.

Head Smut (Sporisorium Reilianum)

Symptoms: appear at the booting stage when the head is replaced by a mass of black spores enclosed in a white fungal membrane. This membrane ruptures on emergence of the head and releases the spores. Partially affected heads are sterile. Head smut is a soil borne disease favoured by cool weather. It may also be introduced on the seed. Control by sowing resistant hybrids. Also avoid sowing susceptible hybrids in cool weather.

Leaf blight (Exserohilum turcicum)

Symptoms: large elliptical spots up to 20 mm wide and 100 mm long, initially water soaked, but drying to straw-coloured spots with red, purple or tan margins depending on the hybrid. Spores produced on leaf spots during moist weather are spread by wind. The fungus survives on undecomposed sorghum residues, volunteer sorghum plants and Johnson grass. Severe disease can cause pinched grain and lower yields. The disease may be serious on susceptible hybrids under humid conditions and in coastal areas. Control by sowing resistant hybrids where the disease may be a problem.

Virus Diseases of Soybeans

Several viruses attack soybeans in Illinois, including soybean mosaic, bean yellow mosaic, tobacco ringspot, cowpea severe mosaic, and bean-pod mottle. The cowpea mosaic and bean-pod mottle viruses have been found only in the southern half of the state. Soybean mosaic virus occurs throughout Illinois with a low incidence present most years. Several strains of the virus have been identified in Illinois.

Soybean Mosaic (SMV)

A plant's reaction to SMV infection depends on the cultivar, the strains of the virus, plant age at time of infection, and environmental conditions (especially temperature). Mosaic is generally a minor disease appearing to a limited extent throughout Illinois. Infected plants are somewhat stunted and bushy, and have distorted leaves. The leaves may be dwarfed, crinkled, or ruffled and are narrower than normal, with margins curling downward.

They may have a yellowish cast and usually show a dark-green, blister-like puckering along the veins. The youngest and most rapidly growing leaves show the most severe symptoms. Nodules on the roots of plants infected with soybean mosaic virus are fewer, smaller, and lighter-weight than those found on healthy plants. Reduced nodulation and leaf distortions are thought to be the primary cause of smaller, lightweight beans in mosaic-infected fields. The pods are often stunted, flattened, or curved with less pubescence; also, they contain fewer seeds. Mosaic-infected plants may produce seed with a black or brown mottling of the seed coat. Badly mottled beans are subject to downgrading, especially in overseas trade, where soybeans are used primarily in food products.

The leaf symptoms of mosaic are sometimes confused with 2,4-D injury. Herbicide damage, however, is much more uniform in a field or area than that caused by mosaic. The injury usually becomes progressively less as the distance from the site of application increases. Herbicide injury is most apparent along sprayed roadsides and near sprayed fields of small grain and corn when the ester form of 2,4-D has been applied on a windy day. Mosaic-infected plants are scattered at random in soybean fields. Some strains of the virus delay the maturity of infected plants, resulting in green plants that stand out in the field after healthy plants have matured. Soybean mosaic is transmitted from one season to the next through seeds.

Virus-infected plants can produce virus-infected seeds that are a mottled dark-brown or black when the plants become infected before flowering. However, seed may be mottled and not carrying the soybean mosaic virus, and virus-infected seeds may not be mottled. SMV-infected seeds may fail to germinate or may produce diseased, spindly seedlings with mottled, crinkled unifoliate leaves.

Infection after flowering ceases may reduce the yield somewhat, but transmission of the virus by seeds to the next generation does not occur. Infection by SMV can predispose some cultivars to infection by pod and stem blight, stem canker, and Phomopsis seed decay resulting in a loss of seed germinability.

The incidence of seed transmission varies from less than 1 percent to 30 percent or more depending on the soybean cultivar, the virus strain, and the time of infection. Transmission of soybean mosaic in the field occurs when certain species of aphids feed on an infected soybean plant and then on a healthy plant nearby.

No aphids colonize soybeans in Illinois, but transient aphid species are plentiful (over 60 species have been identified in soybean fields, of which at least 31 species are able to spread the virus in a nonpersistent manner), especially late in the growing season. Aphids become viruliferous by feeding on infected stems, stem tips, and old or young leaves. Soybean mosaic is a minor disease in full-season soybeans in Illinois because the principal aphid flights occur before planting and after flowering. Soybean mosaic is more prevalent in soybeans planted late, such as after winter wheat.

Control

1. Plant virus-free seed.
2. Do not plant seed produced in late-season plantings.
3. In seed-producing fields, remove (rogue) plants suspected of being virus-infected as soon as they are found.
4. Since SMV is transmitted mechanically, avoid handling or brushing up against infected plants whenever possible.

Although resistance to soybean mosaic virus has been identified (due to one or more dominant genes), resistance to SMV is not yet available in commercial soybean varieties.

Bean Yellow Mosaic (BYMV)

The bean yellow mosaic virus is widely distributed in Illinois, but has not caused evident damage. Affected plants are not noticeably stunted. Usually, the leaves are not distorted and puckered like those affected by soybean mosaic but some strains of BYMV produce severe mottling and crinkling of leaves.

Younger leaves on plants infected with bean yellow mosaic show a scattered yellow mottling or indefinite yellow bands along the major veins. Rusty, dead spots develop in the yellowed areas as the leaves mature. Mixed infections of bean yellow mosaic virus and soybean mosaic virus are common in soybeans.

Bean yellow mosaic virus has a wide host range, is easily transmissible mechanically, and by more than 20 species of aphids in a nonpersistent manner.

Control

Bean yellow mosaic is not seed-transmitted, thus roguing is of no value. No control is warranted. Bud Blight (Tobacco Ringspot Virus [TRSV], Soybean Mosaic Virus [SMV]). Bud blight usually appears first at the border of a field, advancing inward as the season progresses. The disease is common after clover or alfalfa fields have been harvested and when pastures dry up.

Sporadic but serious losses have occurred in some soybean fields, especially in southern and south-central Illinois. Yields may be reduced by 25 to 100 percent. In general, losses are greatest when young plants are infected or when seeds with a high percentage of TRSV are sown. The symptoms vary with the stage of development at which plants become infected. In young plants, the terminal bud turns brown and bends sharply downward forming a crook. Later, other buds become brown. Such buds become dry and brittle and may fall off at the slightest touch.

The leaf just below the tip bud may be bronzed and flecked in a rust colour. A reddish brown discoloration of the pith, especially at the nodes (joints), is often evident inside the stem. Brown streaks are sometimes evident on petioles and along large leaf veins. Plants infected before flowering time are severely dwarfed and produce few or no pods. The pods that do form remain small and are mostly one-seeded. About 90 percent of the seed is infected with virus and will produce virus-infected plants if sown. Infected plants usually remain green until harvested or killed by frost, and they stand out conspicuously in soybean fields at harvest. Plants infected after flowering are much less dwarfed and may produce an almost-normal number of pods. Many pods may drop when they are still small, or may develop dark purple blotches. The seeds on plants affected after flowering do not contain the virus and will produce healthy plants when sown. The virus also significantly reduces root and nodule growth. In instances of severe bud blight, both tobacco ringspot and soybean mosaic viruses have been isolated. Apparently, an interaction between these viruses may induce the leaf to extend its veins beyond the margin of the leaf blade (enations). Many common weeds (e.g., dandelion, wild carrot, purslane, jimsonweed, nettleleaf goosefoot, lamb's-quarters, pigweed, velvetleaf, horse-nettle, sourdock, ragweed, horseweed) and some forage legumes are symptomless hosts of the tobacco ringspot virus. The virus is spread from these plants to soybeans and later from soybean to soybean by unknown vectors. Thrips, grasshoppers, and several other insects have been implicated as well as dagger nematodes.

Control

1. Removing infected plants from fields intended for seed production is of limited value, except in preventing the dissemination of the virus(es) to new areas. Infected

plants produce so little seed and so much of it is lost in threshing that it is of no consequence the next year, and is practically self-eliminating in the second year.

2. Grow 4, 6, or 8 rows of corn, forage sorghum, or other tall-growing crop as a barrier between soybean and clover or alfalfa fields and between soybeans and noncultivated areas, such as roadsides, ditch banks, and pastures.
3. Apply 2,4-D or other herbicide to kill broadleaf weeds in noncrop areas that adjoin soybean fields before the soybean plants emerge or in the fall after the field has been harvested.

Other Virus Diseases of Potential Importance

1. Cowpea Severe Mosaic (CSMV). Symptoms typical of bud blight can also be caused by infection of soybeans with the cowpea severe mosaic virus. The virus, which causes severe mosaic and stunting, is transmitted by seeds in cowpeas but not in soybeans. Numerous leaf-feeding beetles transmit the virus to soybeans, but much less efficiently than to cowpeas. Cowpea severe mosaic was first discovered in Illinois soybeans in 1978 and is not believed to be widespread. Growers who also grow cowpeas should be sure to plant healthy cowpea seeds and to remove virus-infected plants in order to minimize the spread of the virus to soybeans.
2. Bean-pod Mottle (BPMV). The bean-pod mottle virus is present sporadically in Illinois soybeans. It causes a mild and generally unimportant disease (green and yellow mottling of the younger leaves, decreased pod formation and a reduction in seed size, weight, and numbers) in this crop unless the plants are also infected with the soybean mosaic virus. Doubly infected plants are severely stunted with malformed, bronzed leaves, produce few if any pods or seeds, and have a delayed maturity.

Chapter 5

Virus Diseases of Food Graminoid Plants

Maize Disease Management

- Maize is subjected to as many as 112 diseases on a global basis. In India we have records of about 35 of them. These are divisible as
  1. Seed rots and seedling blights
  2. Foliar diseases
  3. Downy mildews
  4. Stalk rots and wilts
  5. Rust
  6. Smuts
  7. Cob and ear rots.
- Blight and stalk rots are the important diseases of maize crop. Blight can be controlled by growing of tolerant / resistant varieties like DHM 103 and Ashwini.
- Stalk rot can be minimized by more frequent irrigations and in case of rainfed crop, soil mulch should be provided preventing loss of moisture through evaporation.

Bacterial Stalk Rot: *Erwinia carotovora*, *Erwinia chrysanthemi*.

Occurrence

- This disease occurs in many states including Andhra Pradesh where high temperatures coupled with high humidity develop during the pre-flowering stage of the crop.

Disease cycle:

- Borer insects play a significant role in initiation of the disease. The organism is soil borne and makes its entry through wounds and injuries on the host surface.
- The organism survives saprophytically on debris of infected materials and serves primary inoculum in the next season.

Epidimeology

- High humidity especially overhead irrigation and waterloging, coupled with high temperature (30°C and above), are conducive for disease development.

Symptoms

- The basal internodes develop soft rot and give a water soaked appearance. A mild sweet fermenting odour accompanies such rotting.
- Leaves some time show signs of wilting or water loss and affected plants within a few days of infection lodge or topple down.
- Ears and shank may also show rot. They fail to develop further and the ears hang down simply from the plant.

Control Measures

- Use of disease resistance varieties, While no absolute resistance has been identified so far, Hybrids Ganga Safed-2, DHM 103, show significantly less disease incidence than other hybrids.
- Avoid waterlogging and poor drainage.

Black Bundle Disease And Late Wilt: *Cephalosporium maydis, Caphalosporium acremonium.*

Epidemeology

- Both the pathogens survive in soil, plant debris and seed. The disease is fovoured by post flowering water stress.

Symptoms

- The disease kills the plant prematurely after flowering. Infected plants do not show symptoms until they reach to tasseling.

- Wilting generally starts from the top leaves, Leaves become dull green, eventually loose colour and become dry.

In advance stages the stalk loses its healthy green colour, lower portions become dry,shrunken with or without wrinklings, hardens and turn purple to dark brown which in more prominent on lower internodes.

- When split open diseased stalks, show brown vascular bundles starting in the underground portion of the roots.
- Diseased plants produce only ears with undeveloped shrunken kernels.
- In severe cases affected plants remain abortive causing 100 per cent loss.
- *Cephalosporium maydis* is primarily soil borne and may infect young maize plants more readily than other plants through roots or mesocotyl. In case of *C. acremonium* only vascular burdles get blackened.

Control Measures

- Use of resistant varieties like DHM 103, DHM 105, Ganga Safed 2.
- Crop sanitation, crop rotations.
- Avoiding water stress at flowering.
- Seed treatment with Thiram or Captan 3g/kg seed.

Charcoal-Rot: Macrophamina Phaseolina

Occurrence

- Charcoal rot is prevalent in comparatively drier maize growing areas, particularly Andhra Pradesh, West Bengal, Bihar, Uttar Pradesh and Delhi.

Epidimeology

- The disease development is maximum during grain filling stage and is favoured by warm temperature (30-40^{0}C) and low soil moisture.
- Charcoal rot is found to be prevalent particularly in *rabi* season when summer temperatures during post flowering period becomes comparatively high (35-45^{0}C).

Disease Cycle

- The pathogen also attack many other hosts, which helps in its perpetuation. Since the fungus is a facultative parasite it is capable of living saprophytically on dead organic tissues, particularly many of its natural hosts producing sclerotial bodies.
- The fungus over winter as a sclerotia in the soil and infects the host at susceptible crop stage through roots and proceed towards stem.

Symptoms

- The characteristic symptoms of the disease become apparent as the plants approach maturity. The disease generally appears early after flowering.
- Plants affected by M. *phaseolina* show evidence of premature ripening. The out sides of lower internodes become straw coloured. The pith becomes badly disintegrated.
- The infected stalks may split longitudinally into a mass of fibres.
- A distinguishing character of the disease is the presence of the small black sclerotia in the pith of the affected stalks. Roots are also invaded and show black sclerotia in the disorganised tissue.

Control Measures

- Regular irrigations particularly during flowering time should be provided.
- Use resistant varieties like DHM 103, Ganga Safed-2 and avoid sowing of suceptable varieties like DHM 105.
- Seed treatment with Carbendazim or Thiram 3g/kg seed is effective.
- Field sanitation, crop rotation should be followed.

Common Rust: Puccinia Sorghi

Occurrence

- This disease is prevalent in cooler parts of the country but its destructiveness is more apparent in rabi maize in Bihar.

- It is very common in Himalayan region during *Kharif* season and in South India during *rabi* season.
- Considerable yield losses were observed in coastal districts in late plantings during rabi.

Epidemiology

- Cold temperature (16-23°C) and higher relative humidity favour the rust development and spread.

Symptoms

- Circular to elongate golden brown or cinnamon brown, powdery, erumpent pustules appear on both leaf surfaces.
- As the crop matures brownish black pustules containing dark thick walled two celled teliospores develop. In severe cases infection spreads to sheaths and other plant parts.

Control Measures

- Plant hybrids like Deccan, Ganga-5, Deccan Hybrid Makka-103 and DHM-1 which are resistant to this disease to minimise the disease intensity.
- Spray Mancozeb 2.5g/lit or Dithane M-45 spray can be taken (0.4%) as soon as first symptoms are observed and it can be repeated at 10 days interval till flowering.

Downy Mildews

- Downy mildews are an exciting group of fungi which attacks many economically important crop plants.
- Ten downy mildew diseases are known on maize caused by two genera viz., *Peronosclerospora* and *Sclerophthera*.
- Out of these, Sorghum Downy Mildew and Brown stripe downy mildew are common in our country.

Sorghum downy mildew: *Perenosclerospora sorghi*.

- Greatest loss results from systemic infection. This most often results in death or barrenness.
- Local infection probably does not cause economic loss in maize.
- Losses have been reported to be as high as 40-60%. Significant loss is usually localized to late planted areas.

Symptoms

- Early maize crops escape infection because by the time conidia are produced they develop resistance on the collateral host susceptibility occurs only upto about 15 days of age.
- In susceptible seedling plants, less than 4 weeks after seedling infection becomes systemic in all growth, subsequent to downward growth of mycelium and colonization of shoot apex (growing point). All *Peronosclerospora* spp. induce both local and systemic infection.
- Malformation of tassels in infected plants.
- Chlorosis, white stripes, stunting with downy fungal growth on both leaf surfaces are the characteristic symptoms.

Life Cycle

- The disease has been know to occur through a collateral host *Heteropogen centortus* on which the fungus perpetuates of the host.
- The role of *Heteropogen* as a collateral host was proved by the cross inoculation tests and occurrence of the disease in the fields adjacent till both conidial and oospores, formed on the collateral host, serve as the primary incolum.
- The breakdown of tissue causes shredding. The oospores either fall to the soil or are wind blown, often within host tissue. They can remain viable in the soil for 5-10 years.
- Conidia are formed at night in large numbers. The optimum temperature for production is 20-23°C.

Brown Stripe Downy Mildew: Sclerophthora Rayssiae

Occurrence

- In India, it has been most severe in Uttar Pradesh, Himachal Pradesh, Southern Rajastan, the Punjab and Hilly parts of West Bengal, especially in areas, which receive 100-200 cm of precipitation.

- A yield loss of 63% was recorded in the Tarai region of Uttar Pradesh

Disease Cycle

- The Oospores present in soil serve as source of primary infection.
- They form sporangia and the infection appears in the form of minute flecks (Characteristics of zoosporic infections) on the leaves, which grow to form the brown streaks.
- Secondary infection is about by sporangia formed on the lower surface of infected leaves.

Crazy Top Downey Mildew: Sclerophthora Macrospora

Symptoms

- Disease symptoms first appear as rolling and twisting of upper leaves before mallformation of tassel.
- The important symptoms of the disease is the partial or complete malformation of the tassel which continues until the tassel resembles a maas of narrow, twisted, leafy structures.
- Affected plants show excessive tillering, stunted with chlorotic stripes on leaves.

Sugarcane Downey Mildew: Peronosclerospora Sacchari

- Characterized by local lesions and systemic infection. Initial leasions are small, round, chlorotic spots on the leaves.
- Systemic symptoms appear as pale yellow to white stripes or streaks at the base of the 3rd to 6th oldest leaves.
- Several streaks may develop on leaf and may extend on the entire leaf.
- Downey growth is seen on the both leaf surfaces. Plants may be distorted with small, poorly filled ears with mis-shapen tassels.

Epidemiology

- High relative humidity with a temperature of 20-25° C favours the disease development. Zinc deficiency predisposes plants to infection

Control measures for downey mildew diseases

- The eradication of collateral and wild hosts near maize field and rouging infected maize plants has been recommended.
- Destruction of plant debris by deep ploughing and other methods.
- Seed treatment with Metalaxyl at 4 g/kg and foliar spray of Mancozeb 2.5 g/l or Metalaxyl MZ at 2g/l is recommended.
- Use of resistant varieties like DMR 1, DMR 5 and Ganga 11.

Head Smut: Sphacelotheca Reiliana

Occurrence

- Reported from Tamilnadu, Andhra Pradesh, Karnataka, Madhya Pradesh, Uttar Pradesh and Punjab

Epidemiology

- The fungus survives in the soil. Soil temperature of 21-28° C with moderate to low soil moisture (15-25%) are optimal for seedling infection.

Symptoms

- Head smut appears when ears and tassels are formed. Floral structures may be partially or completely converted to sori containing masses of brownish black teliospores.
- Tassel infection may be confined to individual spikelets. In such cases floral bracts grow out in to leafy structures, some times in to small shoots.
- Ears infected plants may be smutted or aborted with leafy buds replacing normal ears. Such plants donot produce pollen. infected plants are dwarf with increased tillering.

Control measures

- Field sanitation, crop rotations.
- Seed treatment with Thiram or Captan 3g/kg seed.

Maydis Leaf Blight: Helminthosporium Maydis

Occurrence

- The disease is prevalent in Northern states. It is not a major disease in Andhra Pradesh.

Symptoms

- Lesions on leaf
  Leaves show greyish, tan, parallel straight sided or diamond shaped 1-4 cm long, lesions with buff or brown borders or with prominent colour banding or irregular zonation.
- Symptoms may be confined to leaves or may develop on sheaths, stalks, husks, ears and cobs.
- The lesions longitudinally elongated typically limited to a single inter vascular region, often coalescing to form more extensive dead portions.

Control Measures

- Growing resistant hybrids DHM 103, and DHM 1.
- Use protective fungicides of Carbamate or Mancozeb group.

Maize Mosaic

Symptoms

- Symptoms appear as Chlorotic spots, short lines and broken to nearly continuous Chlorotic stripes covering entire leaf blade.
- Chlorotic stripes and spots can also develop leaf sheaths, stalks and husks. Moderate to severe rosetting of new growth is observed.
- Size of stalk, leaf blades and tassel tend to be normal in late infection.
- Maize mosaic virus is transmitted by Leaf hopper: *Peregrinus maidis*

Control

- Destroy the alternate hosts
- Spray systemic insecticides like Monocrotophos 1.5 ml/l or Dimethoate 2 ml/l for the control of vector.

Maize Dwarf Mosaic: Maize Dwarf Mosaic Virus(MDMV)

Symptoms

- Symptoms first appear on the youngest leaves as irregular, light and dark-green mottle or mosaic.
- Later they develop into narrow sreaks along veins that appear as dark green islands on a clorotic back ground.
- Plants are stunted, with excessive tillering multiple ear shoots and poor seed set.
- The virus is transmitted by Aphids *Rhopolosiphum maidis Myzus persicae, Schizaphis graminicum.*

Control

- Destruction of Johnson grass or other wintering reservoir hosts
- Spray systemic insecticides like Monocrotophos 1.5 ml or Dimethoate 2 ml/l for the control of vector.

Pythium Stalk Rot: *Pythium Aphanidermatum*

Control Measures

- Avoid water logging and close planting.
- Apply Captan (150 gms/100 litres of water) as a soil drench, when the crop is 5-7 week old.
- Hybrids Ganga Safed-2, Hi-starch and DMH-103 should be preferred for avoiding the high incidence of the disease.

Seed Rots and Seedling Disease

- These are incited by species of *Fusarium, Rhizoctonia, Aspergillus, Pythium, Cephalosporium* etc.

Occurrence

- These occur mainly in areas where the soil temperature are low (cooler areas in the Himalayas and in the hill regions of Southern Rajasthan.)

Epidemiology

- The problem becomes severe with the use of old seeds stored under high temperature and humidity and sown during winter.

Symptoms

- Infected seeds when sown, produce thin stands of the crop with gaps, pre-emergence seed rot, post emergence damping off, brown sunken lesions on mesocotyl, collar rot, wilting, toppling of collapsed of seedlings.

Control Measures

- Mechanical sieving or winnowing of seed lots meant for sowing eliminates lightweight, chaffy injured or infected seeds.
- Seed should be treated with the fungicide Thiram or Captan 2 to 2.5 gm/kg of seed by vigorously shaking in a closed metal drum or earthen vessel before sowing.
- Use certified seed.
- Avoid planting in cold, wet or water logged soil.

Smut: *Ustilago Maydis*

Disease Cycle

- The sporidia, produced by germination of teleutospores which infect the soil, cause shoot infection.
- Dikaryotization though not essential for growth in host tissue is however, essential for gall formation.
- The successive generations of teleutospores cause secondary local infections.

Epidemiology

- Disease development is favoured by dry conditions and temparatures between 26-34° C. High nitrogen and injuries increase the potential for smut infection.

Symptoms

- Ground parts are suceptable, particularly young, actively growing meristematic tissues.

- Galls are formed on the ears, axillary buds, tassels, stalks and some times on the leaves ears, axillary buds, tassels, stalks and some times on the leaves causes smut disease in ears and tassels by which they are partially transformed to galls.

Galls on Stalks and Leaves

- Galls are first covered with a glistening, greenish to silvery white tissue. The interior of these galls darkens and turns into masses of powdery, dark olive brown to black spores, except for galls on leaves.
- Plants with galls on lower stalks may be barren and produce small ears. Galls on young seedlings may result in stunting or death of the plants.

Control

- Crop rotation
- Avoiding mechanical injury to plants during cultivation.
- Avoid applying of higher doses of nitrogen.
- Uproot the smut-effected plants.
- Seed treatment with Thiram or Captan 3g/kg seed.

Turcicum Leaf Blight: Helminthosporium Turcicum

- This disease has a world wide distribution. In India the disease is prevalent in the Himalayan region, Peninsular India and the State of Bihar.
- The hot spots being Mandya in Karnataka, Kolhapur in Maharashtra, Karimnagar in Andhra Pradesh and Dholi in Bihar.
- In Andhra Pradesh principal maize growing areas like Warangal, Karimnagar, Medak, Nizambad, Adilabad and Ranga Reddy had recorded more disease in the rabi season.

Occurrence

- This disease is prevalent in the areas where maize is planted in the high lands (more than 1000m altitude), where temperate conditions prevail and in winter plantings in the plain as the cool/moderate conditions favour disease development.

Epidemiology

- The occurrence and damage is severe in areas with heave dew and rainfall with temperature of 18 to 27° C

Symptoms

- Turcicum leaf blight (Northern corn leaf blight) on maize is characterized by long elliptical, grayish green or tan lesions on the leaves measuring 2.5 to 25 cm in length and 4 cm in width.
- Lesions first appear on the lower leaves and increase in size and number until very little living leaf tissue is left. Spores produced on the under surface are arranged in concentric zones resembling a target like pattern when observed against light.
- Crop debris is the usual source of primary inoculum. Once the infection takes place it spreads through wind currents.

Control Measures

- Use of resistance varieties like CM 104, CM 118, CM 121 and DHM 1 is the most effective and economical method of reducing the losses.
- Chemical control as general practice is not recommended to the farmers. However the protection of precious material either experimental or seed production, spraying with Mancozeb @ 0.25 to 0.4% when made at 8-10 days interval is effective.

Maize Dwarf Mosaic, Maize Chlorotic Dwarf Diseases of Corn

Corn Virus Symptom

The two most important viral diseases of corn in Ohio are maize dwarf mosaic (MDM) and maize chlorotic dwarf (MCD). Distribution of these diseases generally reflects the geographic distribution of their overwintering host, johnsongrass. In Ohio, MDM and MCD are serious threats to corn production in the southern half of the state. However, MDM has caused serious economic loss on late planted corn in northern Ohio, especially in the sweet corn growing region along Lake Erie.

Yield losses in fields where the disease is well established may be severe depending on the susceptibility of the corn hybrid being grown. In general, most loss occurs when the plants become infected at the "knee high" stage of development and is minimal on most hybrids if infection occurs after tasseling.

Diagnosis of virus infected plants is difficult based on field symptoms. Samples of plants should be tested in the laboratory to confirm the presence of the virus.

Symptoms

Maize Dwarf Mosaic

MDM symptoms vary with the corn hybrid and the stage of development of the plant at the time of infection. Early infection results in chlorotic spots or flecks that elongate in young leaves in the whorl. Flecks merge into chlorotic streaks along the leaves. These streaks form mosaic or mottled patterns and may turn to a general yellowing as the growing season progresses. Later, plants may have blotches or streaks of red that generally appear after periods of cool (60°F) night temperatures. Infected plants are predisposed to root rot and may be barren.

Maize Chlorotic Dwarf

MCD symptoms include yellowing of youngest leaves in the whorl and a distinct fine yellow striping, or vein clearing of the smallest veins visible between the larger veins. This chlorotic vein clearing of secondary veins is diagnostic for MCD and may be more readily observed on the undersides of infected leaves.

The striping may not be seen in later stages of development due to leaf reddening and general yellowing of the plant. Affected plants may also be stunted due to shortening of the upper internodes.

Disease Cycle

Maize dwarf mosaic virus (MDMV) exists in several strains, the most common being strain A, which infects and overwinters primarily on johnsongrass. Strain B does not infect johnsongrass.

Besides corn and johnsongrass, the MDM virus may infect over 100 wild and cultivated grasses. More than 20 species of aphids can transmit MDMV. An aphid can acquire the virus within a few minutes of feeding on an infected corn or johnsongrass plant.

The aphid then flies or is carried by the wind to other corn plants and inoculates them with the virus. The aphid retains, and is generally able to transmit the virus for 15-30 minutes after acquiring it. The corn leaf aphid and the green peach aphid are common aphid vectors of MDMV in Ohio.

Maize chlorotic dwarf virus (MCDV) is primarily transmitted by the black-faced leafhopper. The leafhopper must feed on an infected plant for several hours to acquire MCDV and then can transmit it for up to 48 hours. Johnsongrass is a known reservoir for MCDV, and its eradication is essential for control of maize chlorotic dwarf.

Control

Grow hybrids tolerant or resistant to maize dwarf mosaic. There is good tolerance and resistance to strain A, but only fair tolerance and no resistance to strain B in dent corn. There is no resistance to maize chlorotic dwarf virus and only fair tolerance. Destroy johnsongrass and other grass hosts, including volunteer corn in areas where corn is to be planted. Best control occurs when all farmers in a community cooperate in eradicating johnsongrass. Plant early, since early planted corn will escape damage because aphid populations do not build up until the plants are past the seedling stage.

Virus Diseases of Corn in Kentucky

The most widely distributed and important virus diseases of corn in Kentucky are maize dwarf mosaic and maize chlorotic dwarf. Maize dwarf mosaic is caused by Maize Dwarf Mosaic Virus (MDMV), while maize chlorotic dwarf is caused by Maize Chlorotic Dwarf Virus (MCDV). These two viruses are usually found together and the disease they cause is often referred to as the "virus complex" of corn. Both field corn and sweet corn are susceptible to the virus complex.

MDMV and MCDV overwinter in rhizomes of johnsongrass and are spread (vectored) to corn by various insects. These viruses can cause sporadic but serious yield loss, depending on the corn hybrid planted, time of infection, amount of johnsongrass, and other factors. MDMV infects over 250 species of grasses, including both annual and perennial species. Sorghum is also an important host of MDMV. The principal hosts of MCDV are corn, johnsongrass, and sorghum, the last two being symptomless hosts.

Symptoms

Symptoms of the corn virus complex can be quite variable under actual field conditions, making diagnosis a problem. Similar symptoms may be caused by other factors such as herbicides, nutrient deficiencies, genetic aberrations, or an adverse environment. Disease diagnosis is even more difficult when plants are infected by both viruses, or when several strains of each virus are present in the field. Consequently, a combination of careful field observations and laboratory tests is the most reliable way to diagnose virus diseases of corn.

Corn plants infected with MDMV often exhibit irregular, light and dark green mottled or mosaic patterns in the leaves, particularly at the base of the youngest leaves. Symptoms of maize dwarf mosaic may be present on all leaves, leaf sheaths, and husks that develop following infection.

In hot weather, the mosaic symptoms may disappear, with young leaves exhibiting only a general yellowing. Infected plants are sometimes stunted, and grain production can be reduced. The symptoms and effects of maize dwarf mosaic are usually most severe in and around areas where rhizome johnsongrass is present. Corn plants infected with MCDV often exhibit shortening of the upper internodes, yellowing of the youngest leaves, reddening of leaf margins, and, occasionally, leaf uttering.

Such symptoms may also be caused by factors other than MCDV. The only diagnostic symptom of maize chlorotic dwarf is a chlorosis of tertiary veins (veinbanding). However, recognition of veinbanding takes considerable training, and definitive diagnosis is best accompanied by laboratory analysis. Severely affected plants may produce little or no grain. As with maize dwarf mosaic, symptoms of maize chlorotic dwarf in corn are typically most severe in and around areas of rhizome johnsongrass. Corn plants infected by both viruses can have a range of symptoms, but affected plants often exhibit severe stunting and reddening and may lack ears.

In johnsongrass, MDMV produces a very faint mosaic symptom which may only be evident at the base of the youngest leaf blades. MCDV produces no symptoms in johnsongrass.

Disease Cycle

Both MDMV and MCDV infect johnsongrass and can survive between corn crops within the underground rhizomes of this weed. Infected johnsongrass rhizomes sprout and produce infected shoots in the spring or early summer. The viruses are then transmitted to corn plants by the feeding activity of several insect vectors (listed below). As the viruses multiply within newly developing tissues, symptoms eventually develop in a susceptible corn plant. Infected corn plants can then serve as a source of virus for new infections to both corn and seedling johnsongrass.

MDMV is transmitted by several aphid species, most commonly the corn leaf aphid, the greenbug, and the green peach aphid. These aphids typically acquire MDMV very quickly during feeding activity, often as quickly as 15 to 60 seconds. MDMV is spread when viruliferous aphids (aphids carrying the virus) fly or are carried by wind to a healthy plant. Viruliferous aphids are capable of infecting a plant after feeding for less than one minute. Outbreaks of maize dwarf mosaic usually are most severe around areas of the field with infestations of infected rhizome johnsongrass. However, some studies have shown that viruliferous aphids may be blown some distance from infected johnsongrass, so the potential exists for outbreaks even in areas of a field with no johnsongrass. MDMV can be seedborne at very low levels, but infected seeds are not considered to be an important source of inoculum in Kentucky.

MCDV is transmitted by several leafhopper species, most commonly the black-faced leafhopper. Usually, several hours of feeding are necessary before leafhoppers acquire or transmit MCDV. Outbreaks of maize chlorotic dwarf typically are limited to areas of the field with infestations of rhizome johnsongrass. MCDV is not known to be seedborne.

Factors Affecting Disease Development

Outbreaks of the corn virus complex are not always predictable and will occur only if the crop is susceptible to infection, the viruses and insect vectors are present, and environmental conditions favour virus spread and symptom development. If one of these factors is missing, then little or no disease will develop. However, when these factors come together, unexpected outbreaks of virus diseases can occur.

Several factors can influence development of virus diseases. Corn hybrids differ greatly in their susceptibility or tolerance to the virus complex, so hybrid selection can influence disease development. Weather and other factors that affect the number and activity of insect vectors can also have an impact on the virus complex.

Late-planted corn is more susceptible to the virus complex. Compared to corn planted on time, late-planted corn is at an earlier stage of crop development during periods of peak activity of insect vectors. Earlier infection of corn usually results in more severe symptoms of the virus complex and greater yield loss. Also, studies by UK entomologists have shown that populations of leafhopper vectors can be higher in late-planted corn.

The amount of johnsongrass present in the field can also affect disease development, as fields highly infested with rhizome johnsongrass usually are highly infested with virus. Even in fields where rhizomes have been eradicated, johnsongrass infestations in the fence rows can serve as a source of virus. This "border effect" is particularly important in small fields or long narrow fields. Serious outbreaks of the virus complex have occurred in some corn fields treated with postemergence grass herbicides. There is some concern that these herbicide applications may occasionally increase the severity of the virus complex. Some research shows increased virus pressure where these herbicides are used. Other research—including our own in 1993—shows no effect of the herbicides.

Disease Management

Like most plant diseases, the corn virus complex is best managed by integrating several control practices. A recommended management program includes using virus-tolerant hybrids, controlling johnsongrass, and avoiding late planting. Insecticides are rarely effective and are not recommended.

Virus-Tolerant Hybrids

A virus-tolerant hybrid of corn is recommended for any field with a significant infestation of rhizome johnsongrass. A number of high-yielding, virus-tolerant hybrids of field corn are available for use on Kentucky farms. With a virus-tolerant hybrid, yield loss will be minimal should environmental conditions favour virus spread and disease development.

Johnsongrass Control

Destroying johnsongrass rhizomes in and around a field eliminates the overwintering host of corn viruses. Therefore, controlling johnsongrass greatly reduces the potential for outbreaks of the corn virus complex.

Timely Planting

Outbreaks of the corn virus complex are most severe in late-planted crops, This is because of increased susceptibility of corn to the viruses, as well as greater activity of insect vectors in late planted corn. Thus, avoid planting corn late to reduce the risk of the disease.

Insecticides

Insecticides are not recommended for use in corn to control the virus complex. Studies have shown that insecticides usually provide no protection against MDMV outbreaks. One reason for this is that aphids can acquire and transmit MDMV within minutes, well before many insecticides can act. Studies with MCDV occasionally have shown some yield benefit using insecticides. However, equal or better disease control can be obtained using virus-tolerant hybrids of field corn without the extra cost of treating with insecticides. Furthermore, MDMV, which is not controlled by the insecticide, is usually present with MCDV.

Virus Diseases of Sweet Corn

Only two virus diseases are currently recognized as infecting sweet corn in New York. They are maize dwarf mosaic virus (MDMV) and maize white line mosaic virus (MWLMV). MDMV has been present in the state since the late 1960s; MWLMV was first recognized in the United States in Tompkins County, New York, in 1979. These two diseases differ greatly in their method of transmission and their locations within the state. Maize dwarf mosaic virus (MDMV) was first reported in the state in 1967 and, at that time, was a serious problem for both field and sweet corn. Today, most field corn hybrids are resistant or tolerant to MDMV, and the genes responsible for resistance in field corn have been transferred into sweet corn progenies.

Because dent corn is the source of resistance for MDMV, the release of high-quality sweet corn varieties has been delayed. In addition to infecting all corn types, MDMV also infects many grass species (annual, winter annual, and perennial).

Although many strains of MDMV have been identified, the two most common strains are called A and B. MDMV-A can infect perennial johnsongrass *(Sorghum halepense),* which is its principal weed source, whereas MDMV-B and the other strains cannot infect johnsongrass. However, other grass species can serve as virus reservoirs for strain B.

Both strains of MDMV are transmitted by several aphid species in a nonpersistent manner, with the corn leaf aphid *(Rhopalosiphum maidis)* being the predominant vector. MDMV remains an important virus disease of sweet corn in key fresh-market producing areas of the state, such as the Hudson Valley and selected upstate locations, and for processing sweet corn in several upstate counties.

MDMV symptoms on sweet corn foliage consist of a mosaic pattern of light and darker green streaks along the veinal and interveinal tissues. Symptoms usually persist, but are most striking at the whorl stages. Mosaic symptoms are also evident on the flag and husk leaves. Inside, the ears will show poor cob and tip fill. Infected susceptible plants are stunted and have reduced yields when compared with resistant plants inoculated at the same time.

Several sources of MDMV inoculum are recognized, but opinions differ as to the importance of these sources. MDMV may be seedborne in corn, but this occurrence is rare and the percentage of infestation is low. A second source of virus is the movement of viruliferous aphids (virus ready for transmission) for long distances on low-level winds often associated with storm fronts.

This source was substantiated in several midwestern states and, therefore, cannot be ruled out. A third virus source is perennial and winter annual grasses that can maintain the virus from one season until the next; the aphids acquire the virus from these sources and transmit it to the susceptible corn crop. This source would appear to be the most plausible and important, but proof is difficult.

Several methods for controlling MDMV are available. Although MDMV-resistant sweet corn varieties are being developed, they currently lack good eating quality, are not suitable for processors, and are not presently available in all maturity groups or colours to span the entire planting season.

Previous work in the Hudson Valley indicated that in most years growers who planted susceptible varieties after June 15 ran the risk of reduced yields from virus infection for corn harvested in September. Thus, planting early to avoid peak aphid flights should be done wherever it is practical. Herbicides should be used to eliminate johnsongrass and other perennial grass hosts bordering fields.

Although insecticides are effective in controlling aphid populations, they cannot prevent the introduction of virus by migrant aphids. Mineral oil sprays have been used successfully in other states to control viruses, including MDMV, transmitted nonpersistently by aphids. However, this technique depends upon special spraying apparatus plus thorough and timely application of oils to a fast growing crop and may not be practical for most growers.

Maize white line mosaic virus (MWLMV) is a relatively new virus disease of corn, first identified in the United States at Ithaca, Tompkins County, New York. Since then the virus has been reported from seven additional states located in New England and the north central region (Ohio, Michigan, and Wisconsin). Since 1979, the virus has been found in 21 counties in New York, where it is more extensively distributed than in any of the other states.

MWLMV is not transmitted mechanically or by aphids or leafhoppers. It is, however, soilborne and appears to be transmitted by a soilborne fungus. Several fungi known to parasitize corn roots act as vectors for plant viruses; the fungi require ample soil moisture to allow zoospores to move in the root zone and make contact with roots of germinating seedlings. Symptoms of MWLMV in corn seedlings appear approximately 1 month after planting into infested soil.

A strong mosaic pattern with short chlorotic white lines 1/ 16 inch wide and up to 3/4 inch long appear in and along the veinal tissue. Some particularly susceptible sweet corn varieties display "goose-necking" of the entire plant.

Severely infected plants will fail to produce ears, whereas ears on other plants are poorly developed and unmarketable. The disease is usually associated with lower, wetter areas of fields, but as fungus/virus-infested soil and debris are redistributed, infected plants may appear in level areas of the field.

Not all plants express symptoms even though virus can be recovered from roots and stems. This suggests that the virus may be more extensively distributed than is now recognized. Because the virus is soilborne and depends upon soil moisture for infection, avoid planting in low, wet areas. Select varieties that are less susceptible; contact your Cooperative Extension agent for an up-to-date list. Because the virus can infect all corn types (sweet, dent, Indian) as well as grasses, infested fields should be rotated out of corn for several years.

Wheat Diseases

In Europe

Cereals are at risk from numerous diseases due to the level of intensification necessary for profitable production since the 1970s. More recently varietal diversification, good plant breeding and the availability of effective fungicides have played a prominent part in cereal disease control. Use of break crops and good rotations are also good cultural control measures. The demise of UK straw burning in the 1980s also increased the importance of good disease control.

Active control measures include use of chemical seed treatments for seed-borne diseases and chemical spray applications for leaf and ear diseases. Development of resistance by diseases to established chemicals has been a problem during the previous 30 years.

Fungicides

Some cereal fungicide groups and examples of active ingredients:

- Benzimidazoles
  - carbendazim.
- Ergosterol biosynthesis inhibitors
  - prochloraz
  - flutriafol
  - tetrachonazole.
- Morpholines
  - fenpropimorph.
- Strobylurines
  - kresoxim-methyl.
- Phthalonitriles
  - chlorothalonil.

Principal Diseases

- Barley yellow dwarf virus, BYDV
- Brown rust *Puccinia recondita*
- Common bunt (aka Covered smut) *Tilletia caries*
- Ergot *Claviceps purpurea*
- Eyespot *Pseudocercosporella herpitrichoides*
- Glume blotch *Septoria nodorum* plus *Septoria tritici*
- Leaf blotch *Septoria tritici*
- Mildew *Erysiphe graminis*
- Seedling blight *Fusarium* spp., *Septoria nodorum*
- Sharp eyespot *Rhizoctonia cerealis*
- Spot blotch *Biplolaris sorokiana*
- Take-all *Gaeumannomyces graminis*
- Tan spot *Pyrenophora tritici-repentis*
- Yellow rust *Puccinia striiformis.*

In the USA

Wheat is subject to more diseases than other grains, and, in some seasons, especially in wet ones, heavier losses are sustained from those diseases than are in other cereal crops. Wheat may suffer from the attack of insects at the root; from blight, which primarily affects the leaf or straw, and ultimately deprives the grain of sufficient nourishment; from mildew on the ear; and from gum of different shades, which lodges on the chaff or cups in which the grain is deposited.

Examples of Wheat Diseases

Bacterial Diseases

- Bacterial leaf blight *Pseudomonas syringae* subsp. *syringae*
- Bacterial sheath rot *Pseudomonas fuscovaginae*
- Basal glume rot *Pseudomonas syringae pv. atrofaciens*
- Black chaff = bacterial streak *Xanthomonas campestris* pv. *translucens*
- Pink seed *Erwinia rhapontici.*

Fungal Diseases

- Alternaria leaf blight *Alternaria triticina*
- Anthracnose *Colletotrichum cereale Manns*
- Ascochyta leaf spot *Ascochyta tritici*
- Black head molds = sooty molds *Alternaria* spp., *Cladosporium* spp.
- Common bunt = stinking smut *T. tritici, T. laevis*
- Downy mildew = crazy top *Sclerophthora macrospora*
- Dwarf bunt *Tilletia controversa*
- Ergot *Claviceps purpurea*
- Foot rot = dryland foot rot *Fusarium* spp.
- Leaf rust = brown rust *Puccinia triticina*
- Leaf and glume blotch, *Stagonospora nodorum* or *Phaeosphaeria nodorum*
- Pink snow mold = Fusarium patch *Microdochium nivale*
- Powdery mildew = *Blumeria graminis*
- Scab = head blight *Fusarium* spp., *Gibberella zeae,*
- Septoria blotch *Septoria tritici* = *Mycospharella graminicola,*

- Smut = Ustilaginomycotina clade of the class Teliomycetae, subphylum Basidiomycota
- Storage moulds *Aspergillus* spp., *Penicillium* spp.
- Wheat mildew
- Wheat leaf rust
- Wheat stem rust.

Nematodes, Parasitic

- Grass cyst nematode *Punctodera punctata*
- Root gall nematode *Subanguina* spp.

Viral Diseases and Virus-like Agents

- Agropyron mosaic genus *Rymovirus*, Agropyron mosaic virus (AgMV)
- Barley stripe mosaic genus *Hordeivirus*, Barley stripe mosaic virus (BSMV)
- Oat sterile dwarf genus *Fijivirus*, Oat sterile dwarf virus (OSDV)
- Tobacco mosaic genus *Tobamovirus*, Tobacco mosaic virus (TMV)
- Wheat dwarf genus *Monogeminivirus*, Wheat dwarf virus (WDV)
- Wheat yellow mosaic Wheat yellow mosaic bymovirus.

Barley Yellow Dwarf Virus

Plants infected at an early growth stage (before the 4 to 5-leaf stage) may be severely stunted and may not head. Plants infected after the boot stage show few or no symptoms and may not suffer any yield losses.

Host Crops

Wheat, barley, oats and other grasses.

Biology

Aphids acquire the virus by feeding on infected plants and spread the virus when they feed on healthy plants. Overwintering of the virus occurs in the southern USA. Aphids carrying the virus are moved north by air currents. The virus is not transmitted by seed or soil. The build-up of aphid populations and extensive spread of this virus disease is favoured by winds from the south and by cool, moist summers (which also favour wheat and grasses).

Infections can occur throughout the season and are most abundant when there are high populations of aphids present. The leaf discolouration symptoms that indicate viral infection develop within about 2 weeks of inoculation at temperatures between 18-21 degrees C (65-70 degrees F). When infections occur at temperatures above 30 degrees C (85 degrees F), the symptoms do not develop.

Symptoms of Damage

Leaf discolouration appears as shades of yellow, red or purple from tip to base and from margin to midrib. Look for yellowed or reddened ßag leaves. Infected plants may be stunted and scattered among healthy plants. Note that these symptoms are sometimes confused with nutritional disorders. Yellowing can be confused with nitrogen deficiency.

Scouting Techniques

Look for stunted plants and leaves that have turned golden yellow from the tips downward.

- Economic Thresholds.
- None Available.
- Control Tips.

Planting date affects the severity of the disease and yield loss. Early planted winter cereals and late planted spring cereals may be subject to high aphid populations at early growth stages, resulting in greater yield loss.

Oat varieties differ in their susceptibility to this disease but there is no effective resistance in any of the cereal crops.

Control volunteer wheat, barley and oats because they may serve as late season hosts to the aphids and virus.

Barley Yellow Dwarf Luteoviruses and other Virus Diseases

Wheat is a natural host for many viruses. Brunt *et al.* (1996) reported 55 viruses to which *Triticum aestivum* is susceptible, and Wiese (1987) described around 30 viruses that naturally infect wheat. This chapter will not describe in detail all wheat viruses, but will give an overview of the most important ones and those that could become a problem in tropical wheat-growing areas. Infection by wheat viruses produces symptoms such as foliar chlorosis (mottle, mosaic, streaks and yellowing), necrosis, stunting and rosetting.

Some of the symptoms, especially foliar discoloration, can easily be confused with nutritional and abiotic disorders. Furthermore, it is common to find several viruses infecting the same plant. This multiple infection makes identification complex, but advances in technology (serology and molecular techniques) have made identification, characterization and classification of viruses easier. However, new viruses are still being described and others are awaiting identification.

Of these, only a few cause extensive damage and have an economic effect on wheat production. These will be described in more detail. Throughout this chapter, the common name and standard acronym that appear in Brunt *et al.* (1996) have been used. Others viruses not mentioned in this chapter have been found occasionally on wheat. For more complete information, refer to Brunt *et al.* (1996) and Wiese (1987).

Barley Yellow Dwarf Luteoviruses

Barley yellow dwarf is the most widely distributed and the most economically important virus disease of wheat. It has been extensively studied, and excellent reviews have been published. Barley yellow dwarf was first reported by Oswald and Houston (1953). It is caused by a group of luteoviruses called barley yellow dwarf luteoviruses (BYDVs).

Barley yellow dwarf luteoviruses are not mechanically transmissible, nor through the seed, but are transmitted by aphids in a persistent, circulative but non-propagative manner. Aphids acquire and transmit BY-DVs while feeding on the phloem sieve tube elements of host plants.

The different viruses that cause barley yellow dwarf disease (BYD) were originally characterized by Rochow (1969) and Rochow and Muller (1971), who identified five strains from New York in the United States based on their transmission phenotypes in an experimental system.

The nomenclature used to identify the five strains of BYDV reflected the species of aphid that most efficiently transmitted that strain and were not intended to imply absolute specificity (Power and Gray, 1995). The strains and their principal vectors are RPV (*Rhopalosiphum padi*), RMV (*R. maidis*), MAV (*Sitobion avenae*), SGV (*Schizaphis graminum*) and PAV (*R. padi, S. avenae* and others).

This nomenclature system has been adopted by all BYD researchers. The five strains are also distinguishable serologically. However, aphid-transmission properties do not always correlate with serotypes. The transmission phenotypes are based on the experimental system, and the five New York BYDV strains are not representative of all strains studied.

For example, *Metopolophium dirhodum* (not included in Rochow's experimental system) has been reported to be an efficient vector of MAV and PAV. This species is also used as the principal vector for MAV-Mexico in BYDV screening trials at the International Maize and Wheat Improvement Centre (CIMMYT).

The International Committee on the Taxonomy of Viruses (ITCV) has divided the BYDVs into two viruses, BYDV (PAV and MAV) and cereal yellow dwarf polero-virus, CYDV (RPV) (D'Arcy *et al.*, 1999), according to the previous grouping based on cytopathology, serology and nucleic acid sequences.

Barley yellow dwarf luteoviruses are restricted to the Poaceae (Gramineae). Cultivated hosts include all the major cereal crops: barley, maize, oat, rice, rye and wheat, as well as many annual and perennial cultivated and wild grasses. BYD has been reported from over 50 countries.

Symptoms caused by BYDVs differ with the host species and cultivar, the age and the physiological condition of the host plant at the time of infection, the strain and the environmental conditions; and they can be easily confused with nutritional and abiotic disorders. In wheat, one symptom includes leaf discoloration from tip to base and from margin to centre, in shades of yellow and sometimes red.

Plants are usually stunted, with a decrease in tiller number and biomass and a weak root system. Suppressed heading, sterility and failure to fill grains occur in the most severe cases.

In the field, symptoms usually appear as yellow or red patches of stunted plants. In aeroponic culture, the root system of BYDV-infected seedlings was initially more severely affected than the shoot; stunting was observed four days after infection in roots and only after 18 days in shoots.

In general, PAV causes severe symptoms, MAV moderately severe and RPV, RMV and SGV produce mild symptoms. However, Chay *et al.* (1996) reported PAV isolates ranging from mild to very severe; an RPV isolate producing corkscrew symptoms was isolated in Mexico and in Ecuador, MAV is known to be severe.

Zhang *et al.* (1983) reported the strain GPV, DAV and GPDAV from China. It seems that some of the Chinese strains have a serological relation to the US isolates (MAV and PAV) but that they differ slightly in their aphid-transmission patterns.

Several strains of BYDV can frequently coexist in the same plant. The resulting symptoms can be more severe when the strains are from different groups, but when from the same group, they may result in the anamelioration of symptoms through the mechanism of cross-protection. Mixed infections can also result in an alteration of the transmission pattern through transcapsidation. Losses in wheat due to BYD can be very serious but differ with the BYDV strains, the growth stage at infection, the wheat varieties and the environmental conditions. Pike (1990) and Lister and Ranieri (1995) compiled data on yield losses from different countries.

Losses of 47 percent and 26 percent have been reported after artificial inoculation with PAV in Kenya and Mexico. Losses of around 11 to 12 percent due to natural infection have been reported in Morocco and in Chile. In Australia, Banks *et al.* (1995a) reported yield losses of about 2.2 tonnes/ha in a susceptible wheat and losses of about 1.1 tonnes/ha in tolerant varieties.

The epidemiology of BYD is influenced by the strains involved, the aphid vectors present in the area, the crop rotation, environmental conditions (temperature and rainfall), the time of sowing and the timing of aphid flights. Barley yellow dwarf alternates from reservoir hosts (grasses, maize, other cereals and volunteer plants) to small grain cereals. The relative importance of primary infection and secondary spread differs from region to region and year to year.

The critical time for BYD control is at an early growth stage. The need for aphid control can either be prophylactic or based on a forecasting system, such as the ones described in Europe by Plumb *et al.* (1986) and Gillet *et al.* (1990). The most commonly used aphicides have been organophosphates or synthetic pyrethroids.

Imidacloprid, an insecticidal seed treatment, decreased BYD infection in certain conditions and is now quite widely used in Europe. Success with biological control has been reported from South America, where *S. avenae* and *M. dirhodum* were controlled through the introduction of Coccinellid predators and Aphelinid and Aphidiid parasites and in New Zealand with the introduction of *Aphidius rhopalosiphi.*

As in most areas, natural enemies limit aphid populations, and it is important to integrate chemical and natural control methods. Adequate cultural practices can be used to limit infection with BYDV. Incorporating resistance or tolerance to BYDVs or their vectors is one of the most promising approaches to control.

Most of the screening for field 'resistance' to BYD has actually been directed to the identification of tolerance. In wheat, sources of tolerance have been reported by several researchers. Tolerance in the variety Anza has been associated with the presence of the gene *Bdv1*, a partially dominant, partially effective gene that induces slow yellowing. However, other genes are involved in tolerance to BYD. The winter wheat germplasm Elmo and Caldwell were released as tolerant to BYDV. Comeau *et al.* (1998) selected a durum wheat line with good tolerance to BYD.

A decrease in virus multiplication has been reported from several wheat relatives, such as *Aegilops*, *Elymus*, *Elytrigia*, *Hordeum*, *Leymus* and *Thinopyrum* (*Agropyron*). Recently, much effort has been directed toward incorporating these alien-derived resistances into wheat. *Thinopyrum intermedium* has been widely used to produce resistant introgressed material, such as the TC lines, Zhong 4 and Zhong 5-derived lines, and most recently L1-derived Ym lines.

The 42 chromosome winter wheat line P29 and spring wheats TC5, TC6 and TC9, as well as the genetic stock Z1, Z2 and Z6 with alien-derived resistance, have been registered. The amphiploid derived from the wheat x *Agrotricum* cross was reported to be immune to BYDV. In TC14, the alien segment is located on 7DL. This line has been incorporated into the CIMMYT breeding programme.

Wheat Streak Mosaic Rymovirus

Wheat streak mosaic rymovirus (WSMV) is a serious pathogen of wheat, particularly in the Great Plains region of the United States. It has been reported from several parts of North America, Jordan, Romania, former Yugoslavia, the Russian Federation, Iran and most recently in Mexico.

It is persistently transmitted by the wheat curl mite, an Eryophid mite, formally identified as *Aceria tulipae* and recently named *Aceria tosichella.* Wheat streak mosaic disease (WSM) is estimated to decrease annual wheat yields by about 2 percent/year in the Great Plains region of the United States. Symptoms differ with wheat cultivar, strain of the virus, time of infection and environmental conditions. Even if winter wheat is infected in the autumn, it rarely shows symptoms until spring. Infected plants are stunted, with mottled green-and yellow-streaked leaves. Symptoms range from mild mosaic to severe chlorosis resulting in the death of tillers, reduced seed set and shrivelled kernels.

The mite relies on wind to move from plant to plant or field to field. Wheat streak mosaic is controlled primarily by breaking the life cycle of the mite vector through elimination of volunteer wheat and grasses and delaying planting. Because corn is a host for both the mite and the virus, it is recommended that wheat not be planted next to corn that has not yet matured. The insecticidal seed-coating imidacloprid has recently been shown to be ineffective against the wheat curl mite and WSMV. The genetic control of WSMV can be achieved either by selecting for resistance to the vector or developing lines resistant to the virus. No resistance to WSMV has been found in wheat, but resistance exists in wheat relatives, such as *T. intermedium.* Wheat-*T. ponticum* (OK65C77-6, OK65C93-8) and wheat-*T. intermedium* (KS93WGRC27) lines have been registered as resistant to WSMV. In KS93WGRC27, the gene *Wsm1* confers resistance to WSMV.

Friebe *et al.* (1996) reported resistance different from *Wsm1* in lines 6957 and 6961 derived from *T. intermedium*. Genes conferring resistance to wheat curl mite colonization have been transferred from *Ae. tauschii*, *Cmc1*, and from *Ag. elongatum*, *Cmc2*. A resistance breaking strain of wheat curl mite has been recently reported.

Soil-borne Wheat Mosaic Furovirus and other Furoviruses

Soil-borne wheat mosaic furovirus (SBW-MV) occurs in China, Japan, Italy, France, Germany, Brazil and the United States. In China, soil-borne cereal viruses (including SBWMV and wheat spindle streak mosaic bymovirus) became a serious agricultural problem in the 1970s when susceptible cultivars were grown continuously on a large scale. Annual cereal yield losses were estimated at 400 000 tonnes.

Soil-borne wheat mosaic furovirus is naturally transmitted by the fungus *Polymyxa graminis* and mechanically transmitted to wheat, barley, rye, *Bromus* and *Chenopodium*. *Polymyxa graminis*, is an obligate parasite and is wide-spread in temperate (United States, New Zealand, Europe, China and Japan) and tropical (West Africa, India and South America) regions.

The virus is carried inside zoospores and persists inside the fungal resting spore for many years. The best conditions for disease transmission (as for vector development) are high soil moisture, humid weather, soil temperature around 15° to 18°C and slightly neutral to alkaline soil. The optimum temperatures for symptom development are 15° to 20°C. Symptoms range from mild green to prominent yellow leaf mosaics.

The new unfolding leaves show mottles and streaks. Stunting can range from moderate to severe, and certain strains of the virus can cause rosetting of plants. The symptoms are more prominent in early spring growth and rarely appear in autumn. Control measures include rotation, delayed sowing, increasing fertilizer application and the use of resistant cultivars.

Application of chemical sterilants and fungicides can help reduce inoculum but are not practical for large-scale field control.

Chen and Wilson (1995) indicate that resistance is controlled by a single locus and that it is dominant. In Italy, the most widely grown bread wheat cultivars are resistant to SBWMV. The durum wheat cultivar Ares showed resistance to SBWMV in northern Italy.

The hard winter wheat germplasm, KS92WGRC21 and KS92WGR22, have been registered as resistant to SBWMV. A potential source of resistance to SBWMV has been reported from the wheat-*T. intermedium* addition line L2. Recently, Aubian wheat mosaic virus, a soil-borne wheat tubular virus which is serologically distinct from other soil-borne wheat viruses, was described in France.

Another tubular virus, which has been described in England, has similarities to the Aubian wheat mosaic virus. In China, Ye *et al.* (1999) reported the presence of a new furovirus, Chinese wheat mosaic virus, which presents difference in serology and nucleotide sequence to SBWMV.

Wheat Spindle Streak Mosaic Bymovirus and Wheat Yellow Mosaic Bymovirus

Wheat spindle streak mosaic bymovirus (WSSMV) and wheat yellow mosaic bymovirus (WYMV), before considered to be strains of the same virus, were shown to be distinct species through genome analysis. A multiplex RT-PCR assay permits the discrimination of these two viruses. Nucleotide and amino acid sequence comparison demonstrated that the European and North American isolates were extremely similar and were WSSMV, while the Chinese isolates were close to the Japanese isolates and were WYMV. Wheat spindle streak mosaic bymovirus was first reported from Canada. Wheat spindle streak mosaic bymovirus or WYMV have been reported from the United States, France, Germany, India, Italy, Zambia, China, Japan, the Democratic People's Republic of Korea and the Republic of Korea. Wheat spindle streak mosaic bymovirus and WYMV are economically important wherever they occur. They are both sap-transmitted to wheat and naturally transmitted through the soil by the fungus *Puccinia graminis*. Since WSSMV and WYMV are transmitted by *P. graminis*, much of what has been said previously for SBWMV applies equally to WSSMV. However, WSS-MV is usually more uniformly distributed in fields than SBWMV. Symptoms develop only in spring when soil temperatures are between 5°C and 15°C.

The virus is usually detected earlier in roots than in leaves. The optimum temperature for the development of *P. graminis* in wheat roots is 15° to 22°C, for transmission of the virus 15°C and for virus development 10°C, perhaps the lowest optimal temperature range of any plant virus disease. Wheat spindle streak mosaic bymovirus and WYMV cause similar symptoms on susceptible plants.

Symptoms vary seasonally and disappear soon after the temperature exceeds 20°C. In early spring, symptoms appear as yellow-green mottling, dashes and streaks. The discontinuous streaks are parallel to the vein and form spindles. In cool temperatures, symptoms are more severe, showing necrosis in the chlorotic spindle. Infected plants can be mildly stunted and with few tillers. Autumn infections are most common, and spring infection results in few symptoms.

Control of both viruses is difficult. Slykhuis (1973) reported that the incidence of WSSMV was suppressed by pre-seeding application of urea (NH_4O_3), or poultry manure, each supplying nitrogen (N) at 500 to 1 800 kg N/ha, but not by cattle manure supplying up to 1 000 kg N/ha. However, these measures cannot be recommended for practical disease control. The only practicable way to reduce yield losses is through resistance to WSSMV and WYMV or their vector. Resistance to WSSMV is seen mainly as a tendency to avoid infection. Some cultivars are resistant to WSSMV, and the resistance appears to be dominant and highly heritable. The cultivars KS92WGR21 and KS92WGR22 have been released as highly resistant to WSSMV.

Brome Mosaic Bromovirus

Brome mosaic bromovirus (BMV) was first reported to cause economic damage in wheat in South Africa and to mimic BYDV infection. It has been reported to naturally infect wheat in the former Soviet Union, former Yugoslavia, Hungary and Brazil, as well as in Canada. Some wheat cultivars are symptomless carriers of the disease. Symptoms on leaves are yellow or white spots and streaks that turn into a yellow mosaic pattern. Infected plants can be slightly stunted and produce shrivelled grains. It is transmitted mechanically to cereals and grasses.

Wheat Dwarf Geminivirus

Wheat dwarf geminivirus (WDV) is an economically important disease on spring and winter wheat in France, Sweden and Eastern Europe. The disease also affects rye and barley. Infected plants develop vein distortions on the underside of leaves and develop fine green to yellow spots. Plants are very stunted, especially if they were infected as seedlings. The main vector is the leafhopper *Psammotettix alienus*. Recent outbreaks have been reported from Sweden.

Barley Stripe Mosaic Hordeivirus

Barley stripe mosaic hordeivirus (BSMV) occurs in North America, Europe, Japan, Australia, the former Soviet Union and China, but is not an economically important disease of wheat. It is mentioned here because it is transmitted through the seed and pollen. Barley stripe mosaic hordeivirus causes yellow to white mottling, spotting and streaking on leaves, severe mosaic, dwarfing, excessive tillering and necrosis. Plants grown from BSMV-infected seeds may show symptoms as early as the second or third leaf stage.

Reoviruses

Wheat is a host for several reoviruses: rice dwarf phytoreovirus (RDV), oat sterile dwarf fijivirus (OSDV), maize rough dwarf fijivirus (MRDV) and rice black-streaked dwarf fiji-virus (RBSDV). Nariño dwarf virus (enanismo, possible oryzavirus) has been associated with strong symptoms in Colombia. Another reovirus (Mal del Rio Cuarto) has been reported to cause extensive damage to wheat and maize in Argentina. Symptoms are deformed leaves, spikes and spikelets, shortened internodes, leaves with serrated borders and sterile spikelets. Mal del Rio Cuarto was classified as a strain of maize rough dwarf fijivirus, but recent molecular data do not support this. Reoviruses are exclusively transmitted by leafhoppers (phytoreoviruses) and planthoppers (fijiviruses), which propagate in the vector and the plant.

Rhabdoviruses

There are several rhabdoviruses that infect wheat, but none of them are important economically. They are transmitted in a propagative manner by the planthoppers. Recent reports mentioned the occurrence of American wheat striate mosaic in the United States and barley yellow striate mosaic rhabdovirus in Turkey.

Newly Described Wheat Viruses

Cereal flame chlorosis was first observed in Canada in 1985 and has only been reported in this country. It is soil-borne and associated with specific double-stranded RNAs. Its spread has been associated with a trend towards more intensive cropping and the presence of certain zoosporic fungi. Disease symptoms include leaf chlorosis in a flame-like pattern, severe stunting and sterility. High plains disease has been recently described as associated with WSMV in wheat and maize in the United States and to cause the aggravation of WSMV symptoms. It is transmitted by the wheat curl mite (*A. tosichella*) and has been associated with the presence of a 32kDa nucleoprotein. Indian peanut clump pecluvirus, highly infectious on graminaceous plants, has been reported on wheat in Africa and India. First classified as furovirus, both peanut clump virus and Indian peanut clump pecluvirus belong now to the new group of pecluviruses.

Oats-Pest management

Leaf Rust

Leaf rust is the most serious problem affecting grazing oats in Queensland. Leaf rust is a disease caused by the fungus *Puccinia coronata avenae*. Growing a leaf rust resistant oat variety is the best solution against the disease. Oat varieties with various levels of leaf rust resistance for Queensland are available. Many oat varieties that were once resistant have become susceptible to this disease. The leaf rust fungus has the ability to mutate into new and virulent forms which overcome those varieties that were once resistant. Leaf rust is sometimes confused with red-tipped leaf which is not a disease but is believed to be a nutritional disorder.

Infection conditions:

- The fungal disease causing this condition survives between cropping seasons on wild oats and volunteer oat plants, which provide a continual supply of spores or inoculum for rust outbreaks each year. The orange spores, seen as pustules on the leaf and leaf sheath, are spread by air.

- Given suitable conditions, it takes one to two weeks for a rust spore to infect and produce more spores on the leaf surface.
- In spring and summer the conditions for rust development are the most favourable with rapid build-up of spores producing serious consequences. In severe outbreak of leaf rust, total loss of the crop can occur.
- Leaf rust can also occur in late autumn in early sown crops (such as February and early March plantings). Early planted crops are usually the worst affected with the epidemic developing rapidly during autumn/early winter.
- Leaf rust is more frequently a problem in warm humid regions such as eastern areas of southern Queensland.

Managing Leaf Rust

- The best solution to leaf rust is to grow resistant varieties.
- Sowing too early (before mid-March) or too late (after June) will increase the risk of young crops becoming infected and provide a new source of inoculum. This cycle can be broken. For instance, by delaying sowing until after mid-March exposure to leaf rust is reduced early in the season and crops are still able to attain adequate growth before winter.
- Losses from leaf rust can also be reduced by grazing or cutting rust infested crops before the disease becomes severe. For instance, if leaf rust is obvious below the top two leaves on each stem, crops should be grazed or cut regardless of growth stage.
- Plant in wider rows to produce an open canopy, and reduce losses from trampling.
- Maintaining good soil and crop nutrition with nitrogen will minimise the affects of leaf rust.
- There are no fungicides registered for control of rust in oats in Queensland.

Stem Rust

Stem rust is caused by a fungus (*Puccinia graminis avenae*) and appears as dark, reddish brown pustules on leaves, leaf sheaths, stems and heads of the plant. The pustules produced are often surrounded by ruptured leaf and stem surfaces. Stem rust can appear early in the season, but tends to be damaging towards spring and summer. It is favoured by warm, moist conditions typical of late spring and early summer. Normally stem rust is widespread when the crop comes to head.

The disease spreads as spores by wind. As in leaf rust, the fungus survives between cropping seasons in wild oats and volunteer oat plants providing a source of inoculum for the new crops. Stem rust is not considered a problem for grazing oats as it occurs late in the growing season.

Stem rust limits oat grain production in Queensland and can cause total loss of the grain. In mild cases the disease produces small and light grains or badly lodged crops. The solution to stem rust in oats is to develop resistant varieties through breeding.

Leaf Spot

Leaf spot (*Drechslera avenae*) appears as small light reddish-brown spots on the leaves and leaf sheaths. It is more common in seedlings and early stages of growth of the oat crop. These spots soon develop into blotches that are oblong to linear with irregular margins. At a later stage, they turn to reddish-brown in colour. In extreme cases, the infected leaf becomes yellow and dries out. The fungus spreads as spores from plant debris. Infection on plant parts is commonly associated with splashes of rain drops. Spores are often lodged in the hulls of the grain, providing a source for infection. Leaf spot could be a problem for grazing oats in some years. It frequently appears on the leaves of infected plants early in the season, although with regrowth its incidence could diminish. With serious infection both the yield and quality of the forage can be affected.

Septoria Blotch

Septoria blotch (*Phaeosphaeria avenaria*) first appears on leaves as small, dark brown spots. At a later stage, these spots develop into larger, circular to oval-shaped blotches, surrounded by light yellow areas. These blotches often join together, producing larger lesions. In extreme cases, the infection can kill the entire leaf, if not a larger part of the leaf.

The infection may spread to the leaf sheaths and stems resulting in lodging. Septoria blotch appears early in the growing season (i.e. autumn) and spreads rapidly in cool, rainy weather. The fungus is transmitted by spores commonly found in plant debris. These spores spread to young seedling through splashes of rain drops. Septoria blotch is not considered a problem for grazing oats. Although this disease shows up early in the growing season, regrowth is generally free of it. It could be a problem for oats grown for hay or grain.

Bacterial Blight

Bacterial blight (*Pseudomonas syringae*) appears on leaves and leaf sheaths as halo blight and stripe blight. Halo blight initially appears as light green water-soaked oval spots (halos) which eventually turn into brown blotches. Stripe blight appears initially on leaf and leaf sheaths as light green spots. Unlike the spots of halo blight, these spots are not surrounded by a pale green margin.

The lesions on the stripe blight eventually lengthen and turn brown, forming water-soaked brown stripes. Blotches produced by both halo and stripe blights often join together to produce large irregular blotches. In both cases, the heavily infected leaves soon turn brown and wither, starting from the leaf tip. At this stage, it is difficult to distinguish the two forms.

The bacteria survive on seeds and plant debris and are often spread by splashes of rain drops, by leaf contact or by insect such as aphids. Bacterial blights could be a problem as it is often widespread early in the season. It could be managed by grazing the crop to remove the infected tissues.

Chapter 6

New and Emerging Plant Viruses

Plant virus diseases, like diseases caused by other pathogens, appear to be proliferating at ever increasing rates. Scientific and popular media abound with terms such as new, emerging, re-emerging, and threatening human, animal, and plant diseases. Is it the result of microscopic agents mounting a concerted attack on higher plants and animals? Is it man's over-populating the planet and his activities that are disrupting natural ecological balances between pathogens and hosts? For eons of time, pathogens and hosts have struggled for survival and equilibria but nowhere in time or space have the opportunities for pathogens and hosts been as fluid as in the last few decades. Levins et al. (1993) identified many conditions that they believe have given rise to new diseases in humans and the same conditions relate to plant virus emergence. Directly and indirectly, by forethought or in hindsight, man is and has been instrumental in the spread and establishment of new disease. Lists of New, Emerging, Re-emerging, and Threatening Plant Diseases have been developed and published in Phytopathology News and included in posters at meetings on several occasions. The original development of such lists came from responses to a survey sent out under the auspices of former APS President Cleo D'Arcy in 1994 for the APS National Plant Pathology Board, headed by Dr. Anne Vidaver.

Pathogen Categories

Pathogens can be classified in four categories: (1) New-pathogens detected within the last five years; (2) Emerging-pathogen incidence has increased within the last 20 years; (3)

Re-emerging-pathogens associated with chemical resistance or changes in management or cultivars, previously controlled infections; and, (4) Threatening-pathogens not reported or limited in distribution in the U.S. A fifth category of 'Chronic/spreading' for pathogens known for longer than 20 years and causing increased concern also is included in the lists.

Goals of the Colloquium

Our goal in preparing for this Colloquium on New and Emerging Plant Viruses has not been to formulate an all-inclusive list of potential new and emerging diseases but to discuss some of the underlying causes for the discovery, development, and understanding of how and why new virus diseases arise, increase in importance, invade new territories, and then wane. We trust this Colloquium will inform, generate interest, answer some questions, and create an awareness of current plant virus problems. Plant viruses have been recognized as plant pathogens for just over 100 years and the number of recognized plant viruses has grown exponentially, in part by the discovery of new virus diseases, in part by the recognition of existing problems as virally caused diseases, in part by the origin of new virus entities and, in part by refining our techniques for separating and identifying the correct causal agents. Viruses require avenues of entry into their plant hosts and arthropods of many types play the role of vectors Nature requires variability for selection to occur. Variability may be driven by the host component as new resistance genes are developed against a population of vectors and viruses, by the vector component as they mutate and adapt to new host systems, or by the virus as mutations and recombinations occur.

Man plays an instrumental role in genetically modifying the host plants, by eliminating weeds and insects with selective pesticides and cultural practices, by rapid dissemination of germplasm to new areas, and by inadvertently or intentionally introducing vectors, viruses, and potential hosts into existing crop systems. These combinations may be detrimental to survival of the viral component or expand opportunities for increase. Examples include: 1) cacao swollen shoot in Africa after cacao was introduced into an environment where the virus had existed in equilibrium with native species; and 2) soybean dwarf in

Northern Japan after soybeans replaced rice over large areas of Hokkaido bringing together a susceptible host with an aphid biotype which preferred it and a virus which had been in equilibrium with native legumes.

The Explosion of the Geminivirus Threat

Not all "new" virus diseases are caused by new viruses or new virus strains but are regarded as new diseases when they appear after a vector species or biotype or a virus is introduced into a new area. One of the virus groups which has literally exploded onto the scene in the last ten years are the Geminiviruses, specifically that group of geminiviruses transmitted by whiteflies. The introduction and spread of a new biotype of *Bemisia tabaci* into new areas has been directly correlated with the increase in new and different geminivirus diseases in tomato, cotton, melons, lettuce, beans, and other crops.

TYLCV is highly virulent to tomato, infects other vegetable and ornamental crops, and is effectively moved short and long distances by seedling transplants. This is a case of an introduction or origination of a new biotype of a vector colonizing an established crop and having a virus introduced which is virulent to the plant host and compatible with the arthropod vector.

Whitefly Transmitted Crinivirus Diseases

Whiteflies, mentioned above as important vectors of geminiviruses, also have been implicated as vectors of a new group of viruses, the 'Criniviruses,' which were previously grouped with the closteroviruses. With the sudden explosion of whiteflies into more temperate regions in the last 10 to 20 years, several new diseases became important yield impacters. Lettuce infectious yellows alone caused losses of lettuce and cucurbits totaling multiple millions in a single season and eliminated fall melons in southern California and parts of Arizona.

Another virus of this group, cucurbit yellow stunting disorder (Figs. 19 and 20), is the most important virus pathogen of cucurbits in southern Europe and the middle East and is a potentially threatening disease for North America.

Figure:

Members of the Criniviruses all induce yellowing symptoms in their plant hosts, are generally phloem-limited, non-mechanically transmissible, and have large ss-RNA genomes. Characterization of one of these viruses, lettuce infectious yellows virus (LIYV), showed that the LIYV genome was composed of two ss-RNAs of 8.1 and 7.2 kb (Fig. 21). RNA 1 encodes for proteins associated with RNA replication and RNA 2 encodes the hallmark closterovirus gene array. LIYV is now recognized as the type member of the genus *Crinivirus* within the family Closteroviridae. Are these new or emerging virus diseases caused by new virus isolates, are they renamed and reclassified because of improvements in molecular technology, or are they the result of a new vector biotype?

Figure:

Factors Contributing to an Increase in New and Emerging Viruses

There are many causes for the apparent increase in new and emerging viruses and in the knowledge gained of viruses not found on the North American continent. In almost all the above scenarios, man plays an important role. As misuse of antibiotics has led to the development of bacteria with multiple resistance, misplaced confidence in our ability to control vectors with pesticides has resulted in multiple chemical resistances (Fig. 22). Germplasm collecting expeditions, international movement of plant material and, lowered trade barriers offers opportunities for unknown viruses to be introduced into new areas.

As we increase the use of transgenic crops carrying viral genes into new vegetative environments, will we continue to increase the rate of new virus evolution, or will we be able to control the viruses now present and create a stable, balanced ecological system?

Figure: *Raspberry bushy dwarf idaeovirus.*

Three components in the virus disease triangle, plant host, arthropod vector, and plant virus all play a role in the rise and fall in occurrence and spread of plant virus diseases. However, super-imposed over all these is the involvement of man as he manipulates the components and the environment in which they interact.

All Plant Viruses have these Common Properties

Some plant viruses have a very limited host range and others attack numerous species.

- Plant viruses can multiply only within living cells.
- Plant viruses usually multiply only within living plant cells, but some may be able to multiply within the bodies of aphids and nematodes.
- A given plant virus may be able to multiply only within the living cells of one species or genera of plants, but some can multiply within the cells of a wide group of plant families.
- We are not sure whether latent plant viruses exist. In general, most organisms harbour both virluent and and latent viruses.
- Plant viruses do not attack animals and vice versa.
- Temperate viruses embedd them selves within the hosts nucleic acid and are transmitted generation to generation just like genes of the host.
- Often viruses reside in their host without causing any disease or symptom. Such latent viruses are undiscovered.

Transmission of Plant Viruses

Plant viruses vary in their mode of transmission. Usually only a single mode of transmission is important, but some viruses are transmitted by more than one mode.

- Aphids or other sap-suckers are the most common mode of transmission
- Nematodes living in the soil transmit some viruses.
- Sap transmission is important for only a few viruses and occurs on cultivators, pruning, hands of workers, and clothing of workers. Important for some potato viruses.
- Pollen transmission from male flower to female occurs for a few viruses. Such viruses are seed borne.
- Most viruses are do not get into the gametes and therefore are not seed-borne.

Detection of Plant Viruses

Until a virus is detected, its presence is not known. Nearly everyone who received an early polio vaccine is now a carrier of Simian Virus 40.(SV40). The monkey cells used to grow that vaccine were infected by SV40 but no one knew SV40 existed.

Many plants are carriers of plant viruses but show no disease. The yield might be reduced but that would not be visbible either.

The methods for detecting plant viruses include:

- Use of antibodies to the virus. In the ELISA assay, plant sap is placed in a plastic tray which contains wells and proteins, including virus, adsorb to the well. Then the tray is washed under running water. Then an antibody to the virus is added. A special enzyme has been bonded to the antibody. Very little of this antibody-enzyme binds to the well because all the plastic of the tray was covered with protein. Wash again. Now a colourless dye complex which the enzyme can split. The dye that is split free has colour. The amount of colour coorelates with the amount of virus present in the donor plant
- Graft a leaf from the suspected plant to an indicator plant. The virus moves into the host plant and causes symptoms in the indicator plant.
- Sap transmission. Place a drop of sap from the suspected plant on an intact leaf of an indicator plant. Add some grit to the sap and rub so the leaf is scratched and the virus, if any, can get into the leaf. If the indicator plant show symptoms then we know the virus came from the donor plant. The indicator plant for sweetpotato viruses is Brazilian Morning glory, tobacco, tomato, lambsquarter, and other plants are used as indicators in the sap and grafting assays.

Creating Virus-free Clones of Plants

It is not possible to cure a plant of viruses, but one of the following methods may give a virus free clone. Since the "virus-free" plant might contain a virus you did not know about, it is proper to call them virus-tested or virus indexed plants.

- Often the terminal bud of a plant is free of virus. Remove the 0.03 to 0.05 mm bud tip and grow it in sterile agar contain sucrose and everything the bud needs. 70 to 99.9% of the buds will rot or fail. If some grow to plants, test that plant bor virus as above and you might find one which is free of virus. Use it to start more plants. Be sure to test the plant for every known virus.

- Some viruses infect the tip also and the above method will not work. In such cases try growing the plant in an incubator that is so hot the plant barely grows. Often you must supply extra carbon dioxide at a very precise level. Then try to start a new plant as above and test it for all the known viruses.
- Most viruses are not seed-borne. Therefore, a new seedling may be free of virus. Test the seedling for all known viruses.

Once you have a virus-indexed plant keep it in a screened cage where it can't be infected via a vector. Alternatively, keep it on sugar agar in a sterile glass jar under fluorescent light at a low temperature where it grows slowly. Use scions from this microplant to grow plants for the field.

Precautions to keep your Plants Virus-free

- viricide your pruning tool between plants. use one volume of Chlorox plus nine volumes of clean water
- Keep the vectors away by screen, culture, or insecticides
- a few viruses can transmitted by clothing brushing against infected plants and then healthy plants. Never fear this transmission is very rare.

Additional Virus Problems

- Different strains of a given virus differ in symptoms and disease virulence just as different humans differ in their criminal activities and kindness.
- Your plant may have an unknown virus or a virus not expected for this plant.
- Some ornamental dwarf plants owe their short stature to a virus(es) and when cured they may grow too tall to be of interest.
- Such virus-dwarfed plants may spread virus to your healthy plants.

- Some plants having varigated foliage owe that beauty to a virus.

Plant Viruses and Viroids

Viruses are pathogens with an extremely narrow host range. Their phylogenetic origin is vague, tough it has always been considered that viruses are genes that became vagrant after having excluded themselves of the host's or a related species' genome. During the last years, an alternative explanation assuming viruses to be by-products of RNA processing has been discussed.

Viruses (Virus particles or virions) are usually units consisting of nucleic acids and coat proteins called capsids. Viroids consist only of RNA, i.e. they contain no protein at all. Except for a few cases, viruses are not surrounded by a membrane. If present, the membrane around a virusparticle-as seen in electron microscopic images-stems usually from the host cell. Viruses have no energy metabolism of their own. Consequently, they cannot perform syntheses and are thus unable to replicate themselves. Depending on their host species, it is distinguished between plant viruses multiplying almost exclusively within plant cells, bacterial viruses (bacteriophages) that depend on living bacteria, and animal viruses.

Picture to the left: Viral membranes. Maturation of the virion (Maus-Friend-leukaemia virus) by budding off the host cell's plasma membrane. Notice the similar structures of the membrane surrounding the virus and the membrane of the host cell (deHARVEN, New York). With plant viruses, the term specificity (or host-specificity) has a very narrow meaning, since no plant virus as such exists. Instead, plant viruses can be grouped in a number of 'varieties'. The tobacco mosaic-virus (TMV), for example, multiplies within *Nicotiana*-species, several other solanaceous plants, and a few species of other plant families. The name of a virus is usually derived from the name of its main host plant. Although with viruses, the term 'species' may not quite correspond to the way it is defined in biological systematics, it is perfectly reasonable and common to use it for viruses, too, since all viruses and viroids contain an original genome with a species-specific information. Its continuity over generations is guaranteed by replication in the host cells. The genetic information of viruses is either encoded by single-stranded RNA (most plant viruses), double-stranded RNA (wound tumour viruses), single-stranded DNA (gemini-viruses) or double-stranded DNA (cauliflower mosaic-virus: CaMV). Based on the shape of the virus particle, it is distinguished between rod-shaped and icosaedrical viruses with a capsid that seems almost spherical.

Tobacco mosaic virus: Electron micrograph of a negatively stained specimen. The relief-like structure is a consequence of image processing. Plant viruses have no specific mechanism for entering the host cell. Cell wall and cuticle are difficult obstacles for them. Plant viruses depend therefore on injuries or on transmission *via* invertebrates. The animal transmitter does in some cases also act as an intermediate host. This means that some plant viruses are able to multiply within animal tissue. Virus diseases of plants are relatively rare. Infection is scarcely strong enough to kill the plant. Monoculture favours spreading, and agricultural losses of profit caused by the potato-X, or the potato-Y viruses, for example, can be significant.

Of many viruses, numerous, considerably differing strains (wildtypes) have been isolated. Among the differences are host range and the degree of virulence. Moreover, viroids were detected. Viroids are small, circular RNA molecules that do not encode proteins themselves. Instead, virions interfere with the transcription of cells due to their similarity with certain areas of recognition of primary transcription products. It seems that viroids prevent the correct cutting out of the introns. They are presumably multiplied with the aid of the cellular DNA-dependent RNA-polymerase II. Viroids occur mainly in warm climates and cause significant loss of profit as the causative agents of the potato disease or the Cadang-Cadang disease of palms. The virus concentration within plant cells is high, although a virus like the TMV does not harm the host seriously. Infected cells contain often voluminous virus crystals. TMV was thus the first virus to be isolated in a pure state and in large amounts. Plants are far from being defenceless against viruses. Only a few virus species are able to penetrate meristematic tissues or to infect a number of successive plant generations (vertical transmission). Hypersensitivity is an effective protection. We know it already from fungus infections.

It is based on the dying of cells in the immediate surrounding of the primary site of infection, thus stopping the spreading of the virus. In the case of *Nicotiana tabacum*, the genetic cause of hypersensitivity has been analyzed. One *Nicotiana* gene product guarantees hypersensitivity against all TMV strains, while another affects only some TMV strains.

The symptom a virus causes at the primary site of infection is called primary symptom. Symptoms caused by its spreading throughout the rest of the plant are called secondary symptoms. Virus infections can usually be recognized by mosaic-like leaf patterns of light and dark green. The infection spreads often over the whole leaf beginning at the leaf veins. Leaves that had been infected during their development are usually deformed or involute.

Frequently, lightened leaf areas, called chloroses, develop around the primary site of infection. Withered areas are called necroses. Chloroses are caused by a breakdown of the chlorophyll resulting in a decreased rate of photosynthesis. Heavy infections are characterized by a complete local loss of chlorophyll.

Affected areas have a yellowish look as only the carotenoids remain. Some TMV strains, for example, can be recognized by yellow leaf areas ("yellow strains"). In nature, such strains are pretty rare since such a strong damaging of the host does also impair the virus' chances of replication and spreading.

Some viruses multiply within the plant without causing symptoms. This phenomenon is called latent infection. In contrast, wound tumour virus cause the development of tumours. The symptoms of most viruses are dependent on both virus and host, and do thus present an important diagnostic feature. G. Samuel proved in 1934 experimentally that the TMV virus uses the plant's vascular system for spreading, actually both the plant's transport to and away. As a result, fully differentiated, old leaves and roots, and young leaves are equally infected.

Since decades now, plant viruses (and especially the TMV) are favourite test objects of pure research due to the large amounts of material extractable from infected plants. The TMV was, too, the first biological electron-microscopic object to be photographed. Isolated TMV-RNA itself is infectious. It can thus be a carrier of genetic information (A. Gierer and G.

Schramm, Tübingen, 1956). Chemical changes of the TMV-RNA (deamination of adenine or cytosine due to treatment with nitrite) showed in addition and for the first time that changes of the RNA are mutagenic, too (K. W. Mundry and A. Gierer, 1958). The changes of these first experiments caused mutations of a coat protein (A. Tsugita and H. Fraenkel-Conrat, Berkeley, 1960; H. G. Wittmann, Tübingen, 1960).

The analysis of a large number of chemically induced mutants (performed by the working groups above (Tsugita and Fraenkel-conrat, Wittmann)) yielded the first results important for the encoding of the genetic code. Wittmann's data proved that the genetic code is not overlapping, and that it is universal in all biological kingdoms: bacteria, plants, and animals. Work on the TMV made plain, too, that the gene-carrying RNA does not simultaneously function as mRNA. mRNA is synthesized within the cell as a separate fraction.

The working group of A. Klug (Medical Research Council, Laboratory of Molecular Biology, Cambridge) studies the assembly of complete virus particles from protein subunits and the incorporation of RNA into the newly formed virus particle since more than two decades. The development of the X-ray analysis (Röntgen structure analysis) and the quicker examination of data due to a new generation of computers allowed the clearing up of the viruses' tertiary protein structure., The substantial advance of the studies reviewed here piloted S. Harrison and his colleagues (Harvard University) in the early eighties to the analysis of spherical viruses. Their studies resulted in the unravelling of the viral tertiary and quaternary structure. The minimum limit of resolution of 2.8 to 2.9 Å was sufficient for the localization of a single amino acid.

Semi-equivalent aggregation was an important discovery in these studies. The coat protein of some viruses is able to aggregate into different forms. The forms differ in their number of symmetry axes and in their molecular assembly. The generation of such aggregates explains an as complex structures as a hollow sphere. The findings of Harrison *et al.* have model character, since complex structures of this type are frequent in nature. The screw-like flagellum of prokaryotes, for example, is an aggregation of a single type of protein (flagellin).

So why have molecular biologists studied bacteriophages instead of plant viruses so intensely and for decades? Here are some plausible answers:

1. The multiplication of bacteriophages is easier to understand. Their hosts, the bacteria (*Eschericia coli* has been studied best) are easy to cultivate, the duration of one generation is short. A complete circle of the phages' multiplication is finished within 15 minutes. The phage / bacterium system is less complex than the virus / plant system. The single parameters are thus easier to measure. Infection with a single phage is sufficient, while in the case of plant viruses often several viruses have to participate.
2. Geneticists studying bacteria and bacteriophages were mainly interested in the (double-stranded) DNA of the phages (RNA-containing phages and single stranded DNA phages exist, too). The information of DNA double-strands can be recombined, the order of its genes is determinable. RNA molecules have no recombination, it is therefore often difficult to say something about the organization of the plant viruses' genomes. Modern genetic engineering, nevertheless, offers possibilities to obtain results. A disadvantage of the phages is the frequently complex structure of their capsids. The coat consists often of more than 20 different proteins.
3. The study of bacteria and their bacteriophages yielded numerous results regarding the course of the infection and the biosynthesis of bacteriophages. Experiments showed the astonishing performances of cells. During the late fifties, for example, mRNA was detected within a phage-infected bacterium. Later on, mRNA was found to be an essential intermediate of every protein biosynthesis.

The study of phytopathogenic viruses raised hopes of obtaining knowledge about the possibilities and the expression of the plant genome, as well as the defence mechanisms of a plant against foreign RNA or DNA.

Viral Diseases

Four virus diseases which occur on citrus in the Lower Rio Grande Valley are exocortis, xyloporosis, tristeza and psorosis. The first three cause rootstock diseases, but psorosis causes bark shelling on the trunk and branches of trees older than 8 to 10 years. These four viruses are bud-transmissible. Certain aphids also are responsible for spread of tristeza virus. Damage caused by virus diseases varies with rootstock-scion combination. Some viruses result in only a slight reduction in growth rate. Others result in yield loss, severe stunting, decline and eventual death.

Because citrus viruses can be transmitted by budding, careful selection of budwood is an effective control method. Tolerant rootstock-scion combinations should be used if virus-free trees are not used. Citrus trees that have been certified to be tristeza-and psorosis-free by the Texas Department of Agriculture should be used as sources of budwood.

Exocortis

This virus causes bark-shelling and stunting of trees on trifoliate orange, trifoliate hybrids and Rangpur lime rootstocks. In the early stages of the disease, gum exudes from pustules at the base of the trunk and may extend from below the soil line to the bud union. New bark forms beneath the pustules, and the outer bark sloughs off, forming the characteristic bark-shelling.

The rate of tree decline varies with tolerance to the exocortis virus. Some affected trees live for quite a number of years, while others die within the first 2 or 3 years. Sour orange and Cleopatra mandarin rootstocks are tolerant to exocortis. Because most of the Valley's citrus is budded on sour orange, exocortis is seldom seen in Texas. Troyer and Carrizo citrange and Swindle citrumelo are rated as moderately resistant. As long as sour orange remains the dominant rootstock in Valley orchards, exocortis should not be a major threat.

Psorosis

Psorosis, also known as scaly bark, has been the most serious virus disease of mature citrus trees in Texas. It spreads chiefly by budding nursery stock with budwood from infected trees. Certain strains of psorosis, however, are capable of being transmitted through seed. Sweet orange, grapefruit and tangerines are severely affected. Most rootstocks used in the Valley are susceptible to the virus; thus, the incidence of psorosis historically has been high.

Several known strains of the psorosis virus have common leaf symptoms but different trunk and branch symptoms. Leaf symptoms are used to identify psorosis-infected nursery stock on young trees. This method of detection has been used effectively in selecting psorosis-free parent trees for budwood. Bark scaling of the trunk and larger branches is typical of psorosis symptoms in citrus trees 8 to 12 years or older. The earliest bark symptom is scaling. As the disease progresses, the tree declines rapidly and eventually becomes unproductive. Psorosis no longer is a major problem in the Valley because most of the infected trees were removed after the 1951 freeze and replaced with psorosis-free trees obtained through the virus-free bud certification program.

Figure: *Bark Scaling from Psorosis.*

Tristeza

This virus has caused serious losses in the citrus-producing areas of South America, South Africa, Australia, California and Florida, but not appreciably in Texas. Because of its severity on the highly susceptible and commonly used sour orange rootstock, it has the potential of being the most devastating disease of Valley citrus. Infected trees found in Texas have been traced to introductions and propagation of infected Meyer lemons and sutsuma mandarin. The black citrus aphid, the most important vector, has not been found in the Valley.

Sour orange and limes, among other rootstocks, are highly susceptible to tristeza. Swingle citrumelo and Troyer and Carrizo citrange are rated resistant. Rough lemon, mandarin, trifoliate orange and its hybrids generally are tolerant to the virus. Because of the danger of tristeza becoming established in Texas, the new citrus rootstocks are commonly investigated in an attempt to find a tolerant rootstock adapted to Valley conditions.

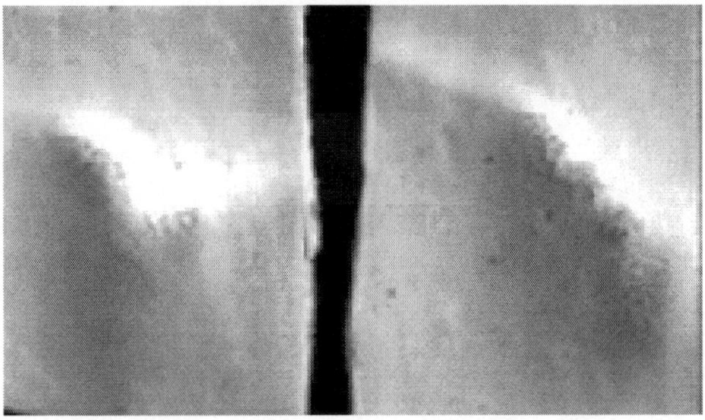

***Figure:** Stem pitting or honeycombing from tristeza found when bark is removed at bud union.*

***Figure:** Leaf flecking and vein clearing on leaf of tristeza-infected tree.*

Symptoms of tristeza-infected trees are similar to those resulting from root injury, including retardation of growth, thinning of foliage, twig dieback and tree collapse under severe conditions. The inside of the bark of tristeza-infected trees shows a honeycomb appearance that can be seen with the naked eye and is used by many as a diagnostic symptom for the disease. In Texas, where the disease has been found with no evidence of insect transmission, no control measure is needed as long as growers plant tristeza-free trees.

Chapter 7

Principles of Diagnosis

The International Workshop on Principles of Diagnosis is an annual event that started in 1989 rooted in the Artificial Intelligence (AI) community. Its current focus is broader covering of a variety of theories, principles, and computational techniques for diagnosis, monitoring, testing, reconfiguration, fault-adaptive control, and repair of complex systems. Application of these theories, principles, and techniques to industry-related disciplines and other domains is amongst the important topics of the workshop. Like the previous workshops in this series, DX-07 encourages the interactions and the exchange of theories, techniques, applications, and experiences amongst researchers and practitioners from different backgrounds: Artificial Intelligence, Control Theory, Systems Engineering, Software Engineering and other related areas, who share an interest in different aspects of diagnosis, and the related fields of testing, reconfiguration, maintenance, prognosis, and fault-adaptive control. DX is a lively forum that has traditionally adopted a single-track program with a limited number of participants in order to promote detailed technical exchange and debate while at the same time making efforts to develop synergistic approaches to solving real-world problems. We welcome papers on topics that are related but not limited to the following:

- Formal theories and computational methods for diagnosis, that include monitoring, detection and isolation, testing, repair and therapy, reconfiguration, fault tolerance, diagnosability analysis, and other related topics.

- Modeling for diagnosis that includes symbolic, numeric, discrete, discrete-event, continuous, hybrid, probabilistic, functional, behavioral, qualitative, abstractions, and approximation methods. Effective modeling approaches for large systems are of particular relevance.
- Computational issues that address combinatorial explosion, use of structural and hierarchical knowledge, focusing strategies, resource-bounded reasoning, real time analysis, and other related topics.
- Diagnosis processes that include strategies for measurement selection, sensor placement, test actions design, active testing, embedded diagnosis systems, preventive diagnosis, fault tolerance strategies, fault-adaptive control, and distributed diagnosis.
- Bridge between DX (AI-based diagnosis methods) and other diagnosis methodologies: FDI, control-based techniques, statistical and probabilistic methods, design, model checking, machine learning, non-monotonic reasoning, planning, execution, real-time languages, software verification and validation, debugging, and hardware testing.
- Real-world applications and integrated systems in a wide range of fields including transportation systems, space and aeronautics, process industries, medical domains, and bioinformatics. Case studies of tech transfer that resulted in success or failure are especially welcome.

Eating Disorders in Adolescents: Principles of Diagnosis and Treatment

Reaffirmed February 2009

Eating disorders are complex illnesses that affect adolescents with increasing frequency. They rank as the third most common chronic illness in adolescent females, with an incidence of up to 5%, a rate that has increased dramatically over the past three decades. Two major subgroups of the disorders are recognized: a restrictive form, in which food intake is severely limited (anorexia nervosa), and a bulimic form, in which binge eating episodes are followed by attempts to minimize the effects of overeating via vomiting, catharsis, exercise or fasting (bulimia nervosa). Both anorexia nervosa and bulimia nervosa can be associated with serious biological, psychological and sociological morbidity, and significant mortality.

Although eating disorders occur most frequently in adolescents, reports in the scientific literature often combine findings from adolescents with those from adults or report exclusively on adult samples. Unique features of adolescents and the developmental process of adolescence are often critical considerations in determining the diagnosis, treatment or outcome of eating disorders.

Consequently, adolescents need to be considered separately and differentiated from adult patients with eating disorders. This position statement addresses the key issues distilled from the scientific literature and represents a consensus of numerous specialists in adolescent medicine regarding the diagnosis and management of adolescents with eating disorders.

Diagnosis

Diagnostic criteria for eating disorders such as described in DSM-IV may not be entirely applicable to adolescents. The wide variability in the rate, timing and magnitude of both height and weight gain during normal puberty; the absence of menstrual periods in early puberty along with the unpredictability of menses soon after menarche; and the lack of psychological awareness regarding abstract concepts (such as self-concept, motivation to lose weight or affective states) owing to normative cognitive development limit the application of those formal diagnostic criteria to adolescents.

In addition, clinical features such as pubertal delay, growth retardation or the impairment of bone mineral acquisition may occur at subclinical levels of eating disorders. The use of strict criteria may preclude the recognition of eating disorders in their early stages and subclinical form (a prerequisite for primary or secondary prevention), and may exclude some adolescents with significantly abnormal eating attitudes and behaviours, such as those who vomit or take laxatives regularly but do not binge. Finally, abnormal eating habits may result in significant impairment in health, even in the absence of fulfilment of formal criteria for an eating disorder. For all of these reasons, it is essential to diagnose eating disorders in adolescents in the context of the multiple and varied aspects of normal pubertal growth, adolescent development and the eventual attainment of a healthy adulthood rather than by merely applying formalized criteria.

Position: In clinical practice, the diagnosis of an eating disorder should be considered in an adolescent patient who engages in potentially unhealthy weight control practices and/ or demonstrates obsessive thinking about food, weight, shape or exercise and not only in one who meets established diagnostic criteria. In such adolescents, an eating disorder should be considered if the teenager fails to attain or maintain a healthy weight, height, body composition or stage of sexual maturation for sex and age.

Medical Complications

No organ system is spared the effects of eating disorders. Although the physical signs and symptoms occurring in a patient are primarily related to the weight control behaviours practised, the health care professional must consider their frequency, intensity and duration, as well as the biological vulnerability conferred by the sexual maturity of the patient. The majority of physical complications in adolescents with an eating disorder appear to improve with nutritional rehabilitation and recovery from the eating disorder, but some may be potentially irreversible. The long term consequences are still to be elucidated.

Medical complications in adolescents that are potentially irreversible include growth retardation if the disorder occurs before closure of the epiphyses; pubertal delay or arrest; and impaired acquisition of peak bone mass during the second decade of life, increasing the risk of osteoporosis in adulthood. These features emphasize the importance of medical management and ongoing monitoring by physicians who understand normal adolescent growth and development.

Just as we endorse early recognition of eating disorders through the use of broad developmentally appropriate criteria, we also endorse early intervention to prevent, limit or ameliorate medical complications, some of which are life-threatening. Adolescents who restrict food intake, vomit, purge or binge in any combination, with or without severe weight loss, require treatment even if they do not meet strict criteria for an eating disorder. Position: Because of the potentially irreversible effects of an eating disorder on physical and emotional growth and development in adolescents, because of the risk of death and because of evidence suggesting improved outcome with early treatment, the threshold for intervention in adolescents should be lower than in adults. Ongoing medical monitoring should continue until the adolescent has demonstrated a return to both medical and psychological health.

Nutritional Disturbances

Nutritional disturbances are a hallmark of eating disorders and are related to the severity and duration of dysfunctional dietary habits. Although abnormalities of minerals, vitamins and trace elements can occur, they generally are not clinically recognized. Deprivation of energy (calories) and protein on the other hand are especially important to identify because these elements are crucial to growth. Moreover, there is evidence that adolescents with eating disorders may be losing critical tissue components, such as muscle mass, body fat and bone mineral, during a phase of growth when dramatic increases in these elements should be occurring. Complete and ongoing assessment of nutritional status is the basis of management of nutritional disturbances in adolescents with eating disorders.

Position: The evaluation and ongoing management of nutritional disturbances in adolescents with eating disorders should take into account the specific nutritional requirements of patients in the context of pubertal development and activity level.

Psychosocial Disturbances

Eating disorders that develop during adolescence interfere with adjustment to pubertal development and the accomplishment of the developmental tasks necessary to become a healthy functioning adult. Social isolation and family conflicts arise at a time when families and peers ought to provide a milieu to support development.

Issues related to self-concept, self-esteem, autonomy, separation from the family, the capacity for intimacy, affective disorders (eg, depression and anxiety) and substance abuse should be addressed in a developmentally appropriate manner. All patients should be evaluated for co-morbid psychiatric illness, including disorders of anxiety, depression, dissociation and behaviour. Because adolescents usually live at home or interact with their families on a daily basis, the role of the family should be explored during both evaluation and treatment.

Position: All adolescents with an eating disorder should be evaluated for co-morbid psychiatric illness. Mental health intervention for adolescents with eating disorders should address not only the psychopathology characteristics of eating disorders, but also the accomplishment of the developmental tasks of adolescence and the specific psychosocial issues central to this age group. For most adolescents, family therapy should be considered as an important part of treatment.

Treatment Guidelines

Because of the complex biopsychosocial aspects of eating disorders in adolescents, the assessment and ongoing management of these conditions appear to be optimal with an interdisciplinary team consisting of professionals from medical, nursing, nutritional and mental health disciplines. Physical and occupational therapy may be useful adjuncts to treatment. Health care providers should have specific experience in treating eating disorders as well as expertise in working with adolescents and their families. They should be knowledgeable about normal adolescent physical and emotional development.

Both in-patient and out-patient treatments need to be available to adolescents with eating disorders. Factors that would justify in-patient treatment include significant malnutrition, physiological or physical evidence of medical compromise (such as vital sign instability, dehydration or electrolyte disturbances) even in the absence of significant weight loss, arrested growth and development, failure of out-patient treatment, acute food refusal, uncontrollable binging, vomiting or purging, family dysfunction that prevents effective treatment, and acute medical or psychiatric emergencies. The goals of treatment are the same in a medical or psychiatric in-patient unit, a day program or out-patient setting: to help the adolescent achieve and maintain both physical and psychological health.

The expertise and dedication of the members of a treatment team who work specifically with adolescents and their families are more important than the particular setting. In fact, traditional settings such as a general psychiatric ward may be less appropriate than an adolescent medical unit, if one of the latter is available. Some evidence suggests that the outcome for patients treated in adolescent medicine units (both out-patient and in-patient) may be better than that of those treated in traditional psychiatric settings with adult patients.

Smooth transition from in-patient to out-patient care can be facilitated by an interdisciplinary team that provides continuity of care in a comprehensive, coordinated, developmentally oriented manner. Health care specialists with an interest in adolescents are familiar with working not only with the patient, but also with the family, school, coaches and other agencies or individuals who are important influences on healthy adolescent development. Given the evidence that eating disorders can be associated with relapse, recurrence, crossover and the later development of other psychiatric disorders, treatment should be of sufficient frequency, intensity and duration to provide effective intervention.

Position: Adolescents with eating disorders require evaluation and treatment focused on biological, psychological and social features of these complex, chronic health conditions. Assessment and ongoing management should be interdisciplinary and is best accomplished by a team consisting of medical, nursing, nutritional and mental health disciplines. Treatment should be provided by health care providers who have expertise in managing adolescent patients with eating disorders and are knowledgeable about normal adolescent physical and psychological development.

Hospitalization of an adolescent with an eating disorder is necessary in the presence of malnutrition, clinical evidence of medical or psychiatric decompensation or failure of out-patient treatment. Ongoing treatment should be delivered with appropriate frequency, intensity and duration.

Barriers to Care

Interdisciplinary treatment of eating disorders can be time-consuming, relatively prolonged and extremely costly. Lack of access to appropriate interdisciplinary teams or insufficient treatment can result in chronicity, social or psychiatric morbidity, and even death. Some provincial plans limit access to private care resources such as nutrition visits or mental health visits. Absent or low reimbursement rates for psychosocial services results in fewer qualified persons being willing to care for teenagers and young adults with eating disorders. Some older adolescents are no longer eligible for treatment or coverage because of provincial medical insurance rules. Thus, withdrawal from treatment can occur at an age when leaving home, unemployment or temporary employment is the norm. Some institutions have age limit policies that negatively affect treatment and limit access to care during the transition from paediatric to adult care. Legislation should provide reimbursement for intervention by multiple disciplines for adolescents with eating disorders. Coverage should ensure that for adolescents, treatment should be dictated by the severity and range of the clinical situation. The promotion of size acceptance and healthy lifestyles, introduction of prevention programs for high risk adolescents, and strategies for early diagnosis and intervention should be encouraged.

Position: Health care reforms should include provisions that address the needs of adolescents with eating disorders and ensure that they not be denied access to care because of absent or inadequate health care coverage.

Critical Care Medicine: Principles of Diagnosis and Management in the Adult

The specialty of Critical Care Medicine has matured from its early stage in which the first textbooks attempted to define a curriculum and establish the legitimacy of a new discipline to its current reality. To affirm this status, Parrillo and Dellinger have assembled an international panel of adult critical care experts and have created a clinically oriented, multidisciplinary and multispecialty review of the subject "aimed primarily at practicing critical care physicians who spend the better part of their time caring for patients in an ICU. Thus the book would be appropriate for critical care internists as well as for surgical or anesthesia critical care specialists. The goal was to produce the acknowledged "best practice" standard in critical care medicine."

The book is organized into 9 parts incorporating 83 chapters that cover not only the wide range of diseases and general procedures that constitute the generally accepted areas of critical care knowledge, but also discuss the other significant issues that confront the practicing physician dealing with end of life, limitation of care, and triage considerations. Administrative considerations are discussed, and the burgeoning responsibilities of critical care practitioners are mentioned. The book addresses the core knowledge requirements through the common technique of addressing disease pathophysiology followed by logical management procedures and best practice guidelines. The parts listed with their integrated chapters create an impressive core curriculum for adult critical care medicine:

I. Critical Care Procedures, Monitoring, and Pharmacology
II. Critical Care Cardiovascular Disease
III. Critical Care Pulmonary Disease
IV. Critical Care Infectious Disease
V. Renal Disease and Metabolic Disorders in the Critically Ill
VI. Neurologic Disease in the Critically Ill
VII. Physical and Toxic Injury in the Critically Ill
VIII. Administrative, Ethical, and Psychological Issues in Care of the Critically Ill
IX. Other Critical Care Disorders and Issues in Care of the Critically Ill·

Previous editions of this text have focused heavily on shock states and resuscitation; these subjects are well covered in this edition, and the expertise of the authors is easily appreciated in Part II, Critical Care Cardiovascular Disease. Chapters on shock, heart disease and specific cardiovascular emergencies are well prepared and informative. The mainstay of a critical care textbook rests on its ability to discuss and provide appropriate instruction in the resuscitation and maintenance of patients requiring mechanical ventilator assistance and other organ support. Complex procedures are well described, but despite the comprehensive nature of this text, practitioners will need to refer to the original citations and subject-specific standard texts, rather than utilizing the descriptions herein as being comprehensive. Certain concepts are ignored; the acid-base chapter is well written and detailed, but space constraints reduce the Stewart discussion of strong ion difference to a paragraph and temperature correction to a sentence. However, the authors obviously recognize the fact and, in keeping with the authoritative nature of the text, provide references that are relevant, comprehensive, and up to date.

Infectious diseases and specific consideration of nosocomial infections are presented, and current prophylactic measures to reduce/prevent such concerns are discussed in detail. Sections on ventilator-acquired pneumonia and catheter-related bloodstream infections are well handled. The section titled "Antibiotic Stewardship" is a good starting point for the remainder of this excellent discussion.

It is important to note that the text incorporates the increasing clinical utilization of ultrasound and echocardiography with chapters detailing critical care applications of transthoracic and transesophageal echocardiography techniques and enhanced diagnostic approaches. Surprisingly, other uses of ultrasound (central venous and arterial catheterization, diagnostic focused assessment with sonography in trauma [FAST] exams, diagnosis of pneumothorax, sinus evaluation, etc.) are not mentioned, and it is hoped that future editions will add this component as the use of bedside sonography continues to expand. Consideration of transcranial Doppler studies is undertaken in three areas: diagnosis of brain death, assessment of cerebral blood flow, and coma. Obviously, extensive supervised and mentored clinical experience is required before any of these techniques can be utilized independently. There is an attempt made to discuss the importance of newer educational concepts, but it is interesting to note that "evidence" and "outcome" are not found in the index. Conversely, good discussions on quality improvement in numerous contexts are apparent, and a brief section on palliative and end-of-life care is presented with useful check lists for assessing patients' needs and, perhaps more importantly, those of their families. However, given the increasing importance of this aspect of critical care and the recognition that it is often the intensivist who first broaches this subject to families, it is anticipated that later editions of this text will pay increasing attention to this important area. A brief introduction to the European-based competency assessment (competency-based training program in ICM (intensive care medicine) for Europe; CoBaTriCE) describes the importance of competency-based evaluations in critical care education. This is consistent with the current focus of the ACGME in defining the 6 core competencies in American Residency education programs.

National quality and/or surgical quality improvement activities are not discussed, and a section on leadership and management skills would be helpful. Increasingly, the critical care physician is asked to manage the important components of systems-based practice that provide the mainstay of current ICU management. Despite the scientific approach to disease management, the attention to clinical pharmacology and physiology and the attention paid to machine and information technology interfaces, the book is lacking in a chapter of nursing care and the importance of this integral interaction especially relevant in sections on sedation and ventilator management.

The editors are true to their Critical Care heritage and have assembled a team of clinician leaders who have created what must be considered the standard textbook for the specialty. It is important to note that Parrillo is the Editor-in-Chief of *Critical Care Medicine* and that he and Dellinger are former Presidents of the Society of Critical Care Medicine. Their credentials to edit such a text are impeccable, and their prestige is evident in the list of contributing authors who joined to create the textbook. As they state in the preface, "our view of critical care medicine is mirrored in the organization of the textbook. Modern critical care is a multidisciplinary specialty that includes much of the knowledge and technology contained in many disciplines represented by the classic organ-based subspecialties of medicine as well as the specialties of surgery and anesthesiology.... Each chapter is designed to provide a comprehensive review of pertinent clinical, diagnostic and management issues.

This is primarily a clinical text, thus the emphasis on considerations important to the practicing critical care physician; also presented, however, are the scientific (physiologic, biochemical and molecular biologic) data pertinent to the pathophysiology and management issues." As in all such endeavors, there are missing pieces and some areas are slighted. However, the question for potential readers is whether the editors have fulfilled their mandate to create a comprehensive, articulate, clinically relevant and authoritative work. In this they have succeeded.

Principles of Diagnosis

General Concepts

The clinical presentation of an infectious disease reflects the interaction between the host and the microorganism. This interaction is affected by the host immune status and microbial virulence factors. Signs and symptoms vary according to the site and severity of infection. Diagnosis requires a composite of information, including history, physical examination, radiographic findings, and laboratory data.

Microbial Causes of Infection

Infections may be caused by bacteria, viruses, fungi, and parasites. The pathogen may be exogenous (acquired from environmental or animal sources or from other persons) or endogenous (from the normal flora).

Specimen Selection, Collection, and Processing

Specimens are selected on the basis of signs and symptoms, should be representative of the disease process, and should be collected before administration of antimicrobial agents. The specimen amount and the rapidity of transport to the laboratory influence the test results.

Microbiologic Examination

Direct Examination and Techniques: Direct examination of specimens reveals gross pathology. Microscopy may identify microorganisms. Immunofluorescence, immuno-peroxidase staining, and other immunoassays may detect specific microbial antigens. Genetic probes identify genus- or species-specific DNA or RNA sequences.

Culture: Isolation of infectious agents frequently requires specialized media. Nonselective (noninhibitory) media permit the growth of many microorganisms. Selective media contain inhibitory substances that permit the isolation of specific types of microorganisms.

Microbial Identification: Colony and cellular morphology may permit preliminary identification. Growth characteristics under various conditions, utilization of carbohydrates and other substrates, enzymatic activity, immunoassays, and genetic probes are also used.

Serodiagnosis: A high or rising titer of specific IgG antibodies or the presence of specific IgM antibodies may suggest or confirm a diagnosis.

Antimicrobial Susceptibility: Microorganisms, particularly bacteria, are tested in vitro to determine whether they are susceptible to antimicrobial agents. Some infectious diseases are distinctive enough to be identified clinically. Most pathogens, however, can cause a wide spectrum of clinical syndromes in humans. Conversely, a single clinical syndrome may result from infection with any one of many pathogens. Influenza virus infection, for example, causes a wide variety of respiratory syndromes that cannot be distinguished clinically from those caused by streptococci, mycoplasmas, or more than 100 other viruses. Most often, therefore, it is necessary to use microbiologic laboratory methods to identify a specific etiologic agent. Diagnostic medical microbiology is the discipline that identifies etiologic agents of disease. The job of the clinical microbiology laboratory is to test specimens from patients for microorganisms that are, or may be, a cause of the illness and to provide information (when appropriate) about the in vitro activity of antimicrobial drugs against the microorganisms identified. The staff of a clinical microbiology laboratory should be qualified to advise the physician as well as process specimens. The physician should supply salient information about the patient, such as age and sex, tentative diagnosis or details of the clinical syndrome, date of onset, significant exposures, prior antibiotic therapy, immunologic status, and underlying conditions. The clinical microbiologist participates in decisions regarding the microbiologic diagnostic studies to be performed, the type and timing of specimens to be collected, and the conditions for their transportation and storage. Above all, the clinical microbiology laboratory, whenever appropriate, should provide an interpretation of laboratory results.

Manifestations of Infection

The manifestations of an infection depend on many factors, including the site of acquisition or entry of the microorganism; organ or system tropisms of the microorganism; microbial virulence; the age, sex, and immunologic status of the patient; underlying diseases or conditions; and the presence of implanted prosthetic devices or materials. The signs and symptoms of infection may be localized, or they may be systemic, with fever, chills, and hypotension. In some instances the manifestations of an infection are sufficiently characteristic to suggest the diagnosis; however, they are often nonspecific.

Specimen Selection, Collection and Processing

Specimens selected for microbiologic examination should reflect the disease process and be collected in sufficient quantity to allow complete microbiologic examination. The number of microorganisms per milliliter of a body fluid or per gram of tissue is highly variable, ranging from less than 1 to 10^8 or 10^{10} colony-forming units (CFU). Swabs, although popular for specimen collection, frequently yield too small a specimen for accurate microbiologic examination and should be used only to collect material from the skin and mucous membranes.

Because skin and mucous membranes have a large and diverse indigenous flora, every effort must be made to minimize specimen contamination during collection. Contamination may be avoided by various means.

The skin can be disinfected before aspirating or incising a lesion. Alternatively, the contaminated area may be bypassed altogether. Examples of such approaches are transtracheal puncture with aspiration of lower respiratory secretions or suprapubic bladder puncture with aspiration of urine.

It is often impossible to collect an uncontaminated specimen, and decontamination procedures, cultures on selective media, or quantitative cultures must be used. Specimens collected by invasive techniques, particularly those obtained intraoperatively, require special attention. Enough tissue must be obtained for both histopathologic and microbiologic examination. Histopathologic examination is used to distinguish neoplastic from inflammatory lesions and acute from chronic inflammations. The type of inflammation present can guide the type of microbiologic examination performed. If, for example, a caseous granuloma is observed histopathologically, microbiologic examination should include cultures for mycobacteria and fungi. The surgeon should obtain several samples for examination from a single large lesion or from each of several smaller lesions. If an abscess is found, the surgeon should collect several milliliters of pus, as well as a portion of the wall of the abscess, for microbiologic examination. Swabs should be kept out of the operating room.

If possible, specimens should be collected before the administration of antibiotics. Above all, close communication between the clinician and the microbiologist is essential to ensure that appropriate specimens are selected and collected and that they are appropriately examined.

Microbiologic Examination

Direct Examination

Direct examination of specimens frequently provides the most rapid indication of microbial infection. A variety of microscopic, immunologic, and hybridization techniques have been developed for rapid diagnosis.

Sensitivity and Specificity

The sensitivity of a technique usually depends on the number of microorganisms in the specimen. Its specificity depends on how morphologically unique a specific microorganism appears microscopically or how specific the antibody or genetic probe is for that genus or species. For example, the sensitivity of Cram stains is such that the observation of two bacteria per oil immersion field (X 1,000) of a Gram-stained smear of uncentrifuged urine is equivalent to the presence of $> 10^5$ CFU/ml of urine. The sensitivity of the Gram-stained smear for detecting Gram-negative coccobacilli in cerebrospinal fluid from children with *Haemophilus influenzae* meningitis is approximately 75 percent because in some patients the number of colony-forming units per milliliter of cerebrospinal fluid is less than 10^4. At least 10^4 CFU of tubercle bacilli per milliliter of sputum must be present to be detected by an acid-fast smear of decontaminated and concentrated sputum.

An increase in the sensitivity of a test is often accompanied by a decrease in specificity. For example, examination of a Gram-stained smear of sputum from a patient with pneumococcal pneumonia is highly sensitive but also highly nonspecific if the criterion for defining a positive test is the presence of any Gram-positive cocci.

If, however, a positive test is defined as the presence of a preponderance of Gram-positive, lancet-shaped diplococci, the test becomes highly specific but has a sensitivity of only about 50 percent. Similar problems related to the number of microorganisms present affect the sensitivity of immunoassays and genetic probes for bacteria, chlamydiae, fungi and viruses. In some instances, the sensitivity of direct examination tests can be improved by collecting a better specimen.

For example, the sensitivity of fluorescent antibody stain for Chlamydia trachomatis is higher when endocervical cells are obtained with a cytobrush than with a swab. The sensitivity may also be affected by the stage of the disease at which the specimen is collected.

For example, the detection of herpes simplex virus by immunofluorescence, immunoassay, or culture is highest when specimens from lesions in the vesicular stage of infection are examined. Finally, sensitivity may be improved through the use of an enrichment or enhancement step in which microbial or genetic replication occurs to the point at which a detection method can be applied.

Techniques

For microscopic examination it is sufficient to have a compound binocular microscope equipped with low-power (1OX), high-power (40X), and oil immersion (1OOX) achromatic objectives, 10X wide-field oculars, a mechanical stage, a substage condenser, and a good light source. For examination of wet-mount preparations, a darkfield condenser or condenser and objectives for phase contrast increases image contrast. An exciter barrier filter, darkfield condenser, and ultraviolet light source are required for fluorescence microscopy.

For immunologic detection of microbial antigens, latex particle agglutination, coagglutination, and enzyme-linked immunosorbent assay (ELISA) are the most frequently used techniques in the clinical laboratory. Antibody to a specific antigen is bound to latex particles or to a heat-killed and treated protein A-rich strain of Staphylococcus aureus to produce agglutination. There are several approaches to ELISA; the one most frequently used for the detection of microbial antigens uses an antigen-specific antibody that is fixed to a solid phase, which may be a latex or metal bead or the inside surface of a well in a plastic tray. The test is then completed by adding a second antigen-specific antibody bound to an enzyme that can react with a substrate to produce a colored product.

Genetic probes are based on the detection of unique nucleotide sequences with the DNA or RNA of a microorganism. Once such a unique nucleotide sequence, which may represent a portion of a virulence gene or of chromosomal DNA, is found, it is isolated and inserted into a cloning vector (plasmid), which is then transformed into Escherichia coli to produce multiple copies of the probe. The sequence is then reisolated from plasmids and labeled with an isotope or substrate for diagnostic use. Hybridization of the sequence with a complementary sequence of DNA or RNA follows cleavage of the double-stranded DNA of the microorganism in the specimen.

The use of molecular technology in the diagnoses of infectious diseases has been further enhanced by the introduction of gene amplication techniques, such as the polymerase chain reaction (PCR) in which DNA polymerase is able to copy a strand of DNA by elongating complementary strands of DNA that have been initiated from a pair of closely spaced oligonucleotide primers. This approach has had major applications in the detection of infections due to microorganisms that are difficult to culture (e.g. the human immunodeficiency virus) or that have not as yet been successfully cultured (e.g. the Whipple's disease bacillus).

Culture

In many instances, the cause of an infection is confirmed by isolating and culturing microorganism either in artificial media or in a living host. Bacteria (including mycobacteria and mycoplasmas) and fungi are cultured in either liquid (broth) or on solid (agar) artificial media. Liquid media provide greater sensitivity for the isolation of small numbers of microorganisms; however, identification of mixed cultures growing in liquid media requires subculture onto solid media so that isolated colonies can be processed separately for identification. Growth in liquid media also cannot ordinarily be quantitated. Solid media, although somewhat less sensitive than liquid media, provide isolated colonies that can be quantified if necessary and identified. Some genera and species can be recognized on the basis of their colony morphologies. In some instances one can take advantage of differential carbohydrate fermentation capabilities of microorganisms by incorporating one or more carbohydrates in the medium along with a suitable pH indicator. Such media are called differential media (e.g., eosin methylene blue or MacConkey agar) and are commonly used to isolate enteric bacilli. Different genera of the Enterobacteriaceae can then be presumptively identified by the colour as well as the morphology of colonies.

Culture media can also be made selective by incorporating compounds such as antimicrobial agents that inhibit the indigenous flora while permitting growth of specific microorganisms resistant to these inhibitors. One such example is Thayer-Martin medium, which is used to isolate *Neisseria gonorrhoeae*. This medium contains vancomycin to inhibit Gram-positive bacteria, colistin to inhibit most Gram-negative bacilli, trimethoprim-sulfamethoxazole to inhibit *Proteus* species and other species that are not inhibited by colistin and anisomycin to inhibit fungi. The pathogenic *Neisseria* species, *N gonorrhoeae* and *N meningitidis*, are ordinarily resistant to the concentrations of these antimicrobial agents in the medium.

The number of bacteria in specimens may be used to define the presence of infection. For example, there may be small numbers ($< 10^3$ CFU/ml) of bacteria in clean-catch, midstream urine specimens from normal, healthy women; with a few exceptions, these represent bacteria that are indigenous to the urethra and periurethral region. Infection of the bladder (cystitis) or kidney (pyelone-phritis) is usually accompanied by bacteriuria of about $> 10^4$ CFU/ml. For this reason, quantitative cultures of urine must always be performed.

For most other specimens a semiquantitative streak method over the agar surface is sufficient. For quantitative cultures, a specific volume of specimen is spread over the agar surface and the number of colonies per milliliter is estimated. For semiquantitative cultures, an unquantitated amount of specimen is applied to the agar and diluted by being streaked out from the inoculation site with a sterile bacteriologic loop. The amount of growth on the agar is then reported semiquantitatively as many, moderate, or few (or 3+, 2+, or 1+), depending on how far out from the inoculum site colonies appear. An organism that grows in all streaked areas would be reported as 3+.

Chlamydiae and viruses are cultured in cell culture systems, but virus isolation occasionally requires inoculation into animals, such as suckling mice, rabbits, guinea pigs, hamsters, or primates. Rickettsiae may be isolated with some difficulty and at some hazard to laboratory workers in animals or embryonated eggs. For this reason, rickettsial infection is usually diagnosed serologically. Some viruses, such as the hepatitis viruses, cannot be isolated in cell culture systems, so that diagnosis of hepatitis virus infection is based on the detection of hepatitis virus antigens or antibodies. Cultures are generally incubated at 35 to 37°C in an atmosphere consisting of air, air supplemented with carbon dioxide (3 to 10 percent), reduced oxygen (microaerophilic conditions), or no oxygen (anaerobic conditions), depending upon requirements of the microorganism. Since clinical specimens from bacterial infections often contain aerobic, facultative anaerobic, and anaerobic bacteria, such specimens are usually inoculated into a variety of general purpose, differential, and selective media, which are then incubated under aerobic and anaerobic conditions.

The duration of incubation of cultures also varies with the growth characteristics of the microorganism. Most aerobic and anaerobic bacteria will grow overnight, whereas some mycobacteria require as many as 6 to 8 weeks.

Microbial Identification

Microbial growth in cultures is demonstrated by the appearance of turbidity, gas formation, or discrete colonies in broth; colonies on agar; cytopathic effects or inclusions in cell cultures; or detection of genus- or species-specific antigens or nucleotide sequences in the specimen, culture medium, or cell culture system.

Identification of bacteria (including mycobacteria) is based on growth characteristics (such as the time required for growth to appear or the atmosphere in which growth occurs), colony and microscopic morphology, and biochemical, physiologic, and, in some instances, antigenic or nucleotide sequence characteristics. The selection and number of tests for bacterial identification depend upon the category of bacteria present (aerobic versus anaerobic, Gram-positive versus Gram-negative, cocci versus bacilli) and the expertise of the microbiologist examining the culture. Gram-positive cocci that grow in air with or without added CO_2 may be identified by a relatively small number of tests.

The identification of most Gram-negative bacilli is far more complex and often requires panels of 20 tests for determining biochemical and physiologic characteristics. The identification of filamentous fungi is based almost entirely on growth characteristics and colony and microscopic morphology. Identification of viruses is usually based on characteristic cytopathic effects in different cell cultures or on the detection of virus- or species-specific antigens or nucleotide sequences.

Interpretation of Culture Results

Some microorganisms, such as *Shigella dysenteriae*, *Mycobacterium tuberculosis, Coccidioides immitis*, and influenza virus, are always considered clinically significant. Others that ordinarily are harmless components of the indigenous flora of the skin and mucous membranes or that are common in the environment may or may not be clinically significant, depending on the specimen source from which they are isolated. For example, coagulase-negative staphylococci are normal inhabitants of the skin, gastrointestinal tract, vagina, urethra, and the upper respiratory tract (i.e., of the nares, oral cavity, and pharynx). Therefore, their isolation from superficial ulcers, wounds, and sputum cannot usually be interpreted as clinically significant.

They do, however, commonly cause infections associated with intravascular devices and implanted prosthetic materials. However, because intravascular devices penetrate the skin and since cultures of an implanted prosthetic device can be made only after incision, the role of coagulase-negative staphylococci in causing infection can usually be surmised only when the microorganism is isolated in large numbers from the surface of an intravascular device, from each of several sites surrounding an implanted prosthetic device, or, in the case of prosthetic valve endocarditis, from several separately collected blood samples.

Another example, *Aspergillus fumigatus,* is widely distributed in nature, the hospital environment, and upper respiratory tract of healthy people but may cause fatal pulmonary infections in leukemia patients or in those who have undergone bone marrow transplantation. The isolation of *A fumigatus* from respiratory secretions is a nonspecific finding, and a definitive diagnosis of invasive aspergillosis requires histologic evidence of tissue invasion.

Physicians must also consider that the composition of microbial species on the skin and mucous membranes may be altered by disease, administration of antibiotics, endotracheal or gastric intubation, and the hospital environment. For example, potentially pathogenic bacteria can often be cultured from the pharynx of seriously ill, debilitated patients in the intensive care unit, but may not cause infection.

Serodiagnosis

Infection may be diagnosed by an antibody response to the infecting microorganism. This approach is especially useful when the suspected microbial agent either cannot be isolated in culture by any known method or can be isolated in culture only with great difficulty. The diagnosis of hepatitis virus and Epstein-Barr virus infections can be made only serologically, since neither can be isolated in any known cell culture system. Although human immunodeficiency virus type 1 (HIV-1) can be isolated in cell cultures, the technique is demanding and requires special containment facilities. HIV-1 infection is usually diagnosed by detection of antibodies to the virus.

The disadvantage of serology as a diagnostic tool is that there is usually a lag between the onset of infection and the development of antibodies to the infecting microorganism. Although IgM antibodies may appear relatively rapidly, it is usually necessary to obtain acute- and convalescent-phase serum samples to look for a rising titer of IgG antibodies to the suspected pathogen.

In some instances the presence of a high antibody titer when the patient is initially seen is diagnostic; often, however, the high titer may reflect a past infection, and the current infection may have an entirely different cause. Another limitation on the use of serology as a diagnostic tool is that immunosuppressed patients may be unable to mount an antibody response.

Antimicrobial Susceptibility

The responsibility of the microbiology laboratory includes not only microbial detection and isolation but also the determination of microbial susceptibility to antimicrobial agents. Many bacteria, in particular, have unpredictable susceptibilities to antimicrobial agents, and their susceptibilities can be measured in vitro to help guide the selection of the most appropriate antimicrobial agent.

Antimicrobial susceptibility tests are performed by either disk diffusion or a dilution method. In the former, a standardized suspension of a particular microorganism is inoculated onto an agar surface to which paper disks containing various antimicrobial agents are applied. Following overnight incubation, any zone diameters of inhibition about the disks are measured and the results are reported as indicating susceptibility or resistance of the microorganism to each antimicrobial agent tested.

An alternative method is to dilute on a log2 scale each antimicrobial agent in broth to provide a range of concentrations and to inoculate each tube or, if a microplate is used, each well containing the antimicrobial agent in broth with a standardized suspension of the microorganism to be tested. The lowest concentration of antimicrobial agent that inhibits the growth of the microorganism is the minimal inhibitory concentration (MIC). The MIC and the zone diameter of inhibition are inversely correlated. In other words, the more susceptible the microorganism is to the antimicrobial agent, the lower the MIC and the larger the zone of inhibition. Conversely, the more resistant the microorganism, the higher the MIC and the smaller the zone of inhibition. The term susceptible means that the microorganism is inhibited by a concentration of antimicrobial agent that can be attained in blood with the normally recommended dose of the antimicrobial agent and implies that an infection caused by this microorganism may be appropriately treated with the antimicrobial agent.

The term resistant indicates that the microorganism is resistant to concentrations of the antimicrobial agent that can be attained with normal doses and implies that an infection caused by this microorganism could not be successfully treated with this antimicrobial agent. (A) Disk diffusion method. (B) Minimum inhibitory concentration (MIC) method. In the example shown, two different microorganisms are tested by both methods against the same antibiotic. The MIC of the antibiotic for the susceptible microorganism is 8 μg/ml. The corresponding disk diffusion test shows a zone of inhibition surrounding the disk. In the second sample, a resistant microorganism is not inhibited by the highest antibiotic concentration tested (MIC > 16 μg/ml) and there is no zone of inhibition surrounding the disk. The diameter of the zone of inhibition is inversely related to the MIC.

Antimicrobial Chemotherapy

Classification

Antimicrobials can be classified by at least three different schemes:

1. Effects on cells
2. Range of activity
3. Sites of activity

This page will examine these different classification schemes and describe several examples of each type of antimicrobial. The mechanisms by which organisms become resistant to these agents will also be discussed. Finally, toxicologic properties of antimicrobial chemotherapy will be described.

Antimicrobial Effects on Cells

Antimicrobials can be divided into two classifications based upon their effects on target cells. Drugs that actually kill microorganisms are termed *bactericidal*. Drugs that only inhibit the growth of microorganisms are termed *bacteriostatic*. The decision to use a bactericidal or bacteriostatic drug to treat infection depends entirely upon the type of infection. For example, bactericidal drugs will only kill cells that are actively growing. Bacteriostatic drugs, in comparison, will only inhibit the growth of cells; ultimate elimination of the organisms is dependent upon host phagocytic activity. Some examples of bactericidal and bacteriostatic drugs are listed below.

Range of Activity

Antimicrobials can also be classified by their range of activity. In general, five classifications can be described. The first of these is termed *narrow spectrum*. Narrow spectrum drugs, as the name implies, are only active against a relatively small number of organisms. In general, narrow spectrum antibiotics are effective against Gram-positive organisms. The second classification is termed *moderate spectrum*. These drugs are generally effective against the Gram-positives and most systemic, enteric and urinary tract Gram-negative pathogens. The beta-lactam antibiotics (penicillin, ampicillin, cephalosporins, etc.) belong in a third classification, *narrow and moderate spectrum* because some members are only effective against Gram-positive organisms while other members can also kill certain Gram-negative bacteria. A fourth classification is termed *broad spectrum*. These drugs are effective against all prokaryotes with two exceptions: *Mycobacteria* and *Pseudomonas*. The fifth group includes those drugs that are effective against *Mycobacteria*.

Antimicrobial Chemotherapy for Legionnaires Disease

Paul H. Edelstein, MD + Author Affiliations

University of Pennsylvania Medical Centre; Philadelphia, PA 19104-4283 Note: Dr. Edelstein has received funding for research and related academic purposes from manufacturers of both macrolide and quinolone antibiotics. Acknowledgment: The author thanks Richard D. Meyer for reviewing the manuscript and suggesting changes. Requests for Reprints: Paul H. Edelstein, MD, Clinical Microbiology Laboratory, 4 Gates, University of Pennsylvania Medical Centre, 3400 Spruce

Street, Philadelphia, PA 19104-4283; e-mail, phe@mail.med.upenn.edu. Erythromycin became the de facto standard of treatment for Legionnaires disease soon after the first recognized epidemic of this disease occurred 22 years ago. However, it is now time for a change in therapy because of increasing evidence of the superiority of newer antimicrobial agents. The conclusion that erythromycin is the drug of choice for Legionnaires disease was based on unavoidably imperfect retrospective analysis of treatment outcomes from the epidemic, which showed the lowest overall fatality rates with either erythromycin or tetracycline therapy.

No prospective comparative studies with adequate statistical power of the treatment of Legionnaires disease have been published. Because of the relative rarity of Legionnaires disease and its widely varying outcomes, it is unlikely that an adequate comparative clinical trial of the disease will ever be completed; such a study would need to have 300 to 900 patients with the disease in each experimental group to have adequate statistical power. This means that decisions about new therapy must be based on laboratory experiments and limited clinical data.

Case-fatality rates for treated Legionnaires disease depend on various factors, but the chief determinants seem to be 1) the premorbid immune and general health status of the patient and 2) whether the disease is community-acquired or nosocomial. Higher fatality rates for nosocomial infection are probably due to host-related rather than organism-related factors. Previously healthy patients with community-acquired Legionnaires disease who do not require hospitalization respond well to erythromycin therapy about 95% to 99% of the time. Quite a different picture emerges for immunocompromised patients with Legionnaires disease or for patients with nosocomial Legionnaires disease. Fatality rates of up to 50% have been reported for treated patients in these categories, with a very wide range; fatality rates as high as 80% have been found in untreated patients. Prolonged illness before therapy and respiratory failure caused by pneumonia are especially poor prognostic factors. Relapse after cessation of erythromycin therapy can be a serious problem in some immunosuppressed patients and can require prolonged and repeated courses of therapy. The occasional ineffectiveness of erythromycin therapy was recently highlighted in a report of fatal *Legionella micdadei* pneumonia in an immunocompromised patient who died despite receiving multiple courses of erythromycin therapy.

In addition to giving rise to potentially poor outcomes in patients with severe Legionnaires disease, erythromycin causes numerous unpleasant side effects. These include high rates of phlebitis and infusion pain with intravenous administration and frequent gastrointestinal problems with oral administration. Large intravenous infusion volumes are often required to increase patient tolerance, and this can exacerbate fluid management problems in critically ill patients. Other possible problems include cardiac arrhythmias and conduction abnormalities, hepatitis, hearing loss, interference with the metabolism of numerous drugs, and poor patient tolerance overall. Of adults who receive outpatient therapy with erythromycin, 12% to 33% do not complete therapy because of intolerance; this contrasts with the much lower dropout rates for most other antibiotics.

Because of adverse effects, hospitalized patients receiving intravenous erythromycin may have significantly increased lengths of stay and increased costs of hospitalization compared with patients receiving other agents. In some outpatients, the frequency of adverse effects may mean that the long courses of therapy required to cure Legionnaires disease are not completed, resulting in relapse. If valid comparative clinical studies cannot be done, how can the value of new drugs be assessed? Luckily, certain types of laboratory studies have correlated very well with clinical response to infection. For example, animal models and intracellular infection models correctly predict that erythromycin is clinically effective and that cefoxitin is not. The success of these correlations is based on the intracellular location of *L. pneumophila* and the ability of drugs to achieve high, bioactive, intracellular concentrations. It is less certain that an antimicrobial agent that is highly active in such experiments is more effective clinically than a drug (such as erythromycin) that is marginally effective in laboratory tests. The lack of controlled clinical trials means that this question is not easily answered. One retrospective study showed that pefloxacin is more active clinically than erythromycin; this finding correlates with the superior laboratory activity of pefloxacin.

In intracellular models of *L. pneumophila* infection, bacterial growth is inhibited by erythromycin and rifampin but recurs when the drugs are removed from the cells. Thus, these drugs differ from many newer fluoroquinolones, which can kill most, if not all, of the bacteria present in the infected cells and can be removed from the cells without causing rapid bacterial regrowth.

Against intracellular *L. pneumophila*, levofloxacin, sparfloxacin, and trovafloxacin are the most active fluoroquinolone antimicrobial agents in the laboratory that are licensed by the U.S. Food and Drug Administration (FDA). Azithromycin is also more active than erythromycin against intracellular *L. pneumophila*.

Studies of animal models of *L. pneumophila* pneumonia confirm the superior activity of most fluoroquinolone antimicrobial agents and azithromycin compared with erythromycin. In these experiments, a single low dose of azithromycin not only is curative but also markedly reduces lung concentrations of *L. pneumophila*. In contrast, multiple, high doses of erythromycin are needed to achieve a similar cure rate without the benefit of rapid lung sterilization.

Extremely low dosages of trovafloxacin or sparfloxacin also have had curative and sterilizing effects in animal studies. The failure of erythromycin treatment to completely eradicate *L. pneumophila* from guinea pigs with experimental pneumonia correlates with disease relapse in some heavily immunosuppressed patients after cessation of erythromycin therapy.

Uncontrolled clinical trials show that many new fluoroquinolone antimicrobial agents effectively treat both community-acquired and nosocomial Legionnaires disease; these agents include ciprofloxacin, pefloxacin, ofloxacin, levofloxacin, and trovafloxacin (trovafloxacin has been studied against community-acquired disease only). Similar clinical data exist for azithromycin.

On the other hand, treatment failures have been reported for these newer antimicrobial agents, as they have for erythromycin, and use of these agents is not a guarantee of therapeutic success. Laboratory evidence suggests that the most active fluoroquinolone antimicrobial agents are trovafloxacin, sparfloxacin, and pefloxacin. Levofloxacin, ofloxacin, and probably grepafloxacin possess intermediate but still high activity. Ciprofloxacin is slightly less active than the other compounds but is still more active than erythromycin. Azithromycin is substantially more active in animal experiments than is clarithromycin. Azithromycin (intravenous formulation), levofloxacin, and trovafloxacin are currently licensed by the FDA for the treatment of Legionnaires disease, as are erythromycin and dirithromycin.

In my opinion, azithromycin or one of the more active fluoroquinolones should be used in preference to erythromycin for the treatment of Legionnaires disease in hospitalized and immunocompromised patients. The potentially high mortality and morbidity rates of the disease in such patients, combined with the experimental evidence discussed above, is reason enough to use more active antimicrobial agents in this setting. In addition to producing a potentially better outcome, these agents will often improve patient convenience as a result of fewer side effects and shorter durations of therapy. Adding rifampin to the above therapies probably provides little further benefit; no convincing laboratory data show that adding rifampin to fluoroquinolone or more active macrolide therapy improves bacterial killing. In our institution, levofloxacin (500 mg/d for 7 to 10 days) or azithromycin (500 mg/d for 3 days; 500 mg on day 1 and 250 mg/d for 4 more days; or 500 mg/d for 5 days) are recommended for the treatment of inpatients with Legionnaires disease. For nonimmunocompromised outpatients with community-acquired Legionnaires disease, erythromycin or tetracycline therapy remain perfectly acceptable. These drugs are much less expensive than the newer agents, and treatment outcomes in this population are likely to be the same with new and old therapies. Ten to 14 days of treatment with erythromycin or doxycycline should be sufficient for such patients.

Chapter 8

Controlling Diseases in Plants

Diseases of vegetables grown in the home garden may reduce both the yield and quality of vegetables. Controlling such diseases often determines success or failure and adds immeasurably to the pleasure derived from the garden. Diseases of vegetables are caused by microorganisms, including fungi, bacteria (including mycoplasmas and spiroplasmas), viruses, and nematodes.

These organisms are spread by wind-blown or water-splashed spores, insects, infected seeds or transplants, by the movement of infested soil, and by humans handling wet plants. Vegetable diseases can be successfully controlled using an integrated disease control program. Such a program involves the use of crop rotation, sanitation, disease-resistant or tolerant varieties, disease-free seeds or transplants, sound horticultural practices, balanced soil fertility (based on soil tests), and proper and timely applications of pesticides. There is no single control practice that can control all of the diseases of any vegetable crop. Therefore, several measures must be used to achieve satisfactory control.

Disease Control Before the Garden is Planted

Choose and prepare the site: The site for the vegetable garden should be well-drained, have adequate sunlight (8 or 10 hours of direct sunshine per day), and good air circulation. The soil should have adequate soil fertility and the proper pH (5.5 to 7.0). Information on garden location, soil fertility, and soil pH can be found in University of Illinois Extension Circular 1331.

Midwest: If the area has produced garden vegetables for several years, sanitation and rotation practices will reduce the risk of disease-causing organisms that have survived from preceding crops. Removing or turning under dead and diseased plant material in the fall is important for disease control. Crop residues from healthy plants can be composted and returned to the garden. Disease-causing fungi and bacteria can survive composting unless the decomposition of the crop residue is complete, a situation that seldom occurs. Therefore, we suggest that no diseased plant parts be placed in the compost pile. Some disease organisms will even survive complete composting, including corn smut; clubroot of cabbage, broccoli, cauliflower and other crucifers; Verticillium wilt of potato, tomato, pepper, and eggplant; and root-knot nematodes. Turning (plowing or tilling) under crop residues in the fall helps prevent the over-wintering of many disease-causing organisms. Besides destroying the "winter home" of various organisms, this practice also helps control the insects that transmit certain diseases. Also, crop debris that is turned under improves soil tilth and makes spring garden work easier.

Crop rotation: Crop rotation is an important control measure since many disease-causing organisms only attack related plants in the same family. Unrelated plants do not serve as hosts on which these organisms can multiply. Therefore, if possible, avoid planting any vegetables within each of the following groups in the same location more than once every three or four years.

Cole crops–(cabbage family) broccoli, Brussels sprout, cabbage, cauliflower, Chinese cabbage, mustard greens, kohlrabi, radish, rutabaga, and turnip. Cucurbits–(cucumber family) cucumber, gourds, muskmelon, pumpkin, squash (all types), and watermelon.

Solanaceous crops–(tomato family) eggplant, husk tomato (ground cherry), potato, pepper, tomato. Chenopodiaceous crops–beet (red and sugar) and spinach.

Amaryllidaceae–(onion family) chives, leeks, onions, shallots, and garlic.

Legumes–pea, beans (all types). Umbelliferous crops–carrot, parsnip, celery. For example, cabbage, broccoli, radish, or turnips should not be planted in the same location for two successive years. Cabbage could be followed with beans, a vine crop such as cucumber, or with sweet corn.

Diseases affecting beans, peas, sweet corn, carrot, lettuce, spinach, rhubarb, okra, edible soybeans, and similar plants are usually specific to only one of these crops, and generally will not infect others or members of the major vegetable crop groups.

Choose disease-resistant varieties: The use of well-adapted, disease-resistant varieties is the simplest and most efficient method of controlling many diseases. Lists of disease-resistant vegetables recommended for use by Illinois gardeners are given in Circular 1331 (Vegetable Gardening in the Midwest. Additional information on disease resistance is also given in the following paragraphs. A listing of varieties and their resistance can be found in Circular 1363, Midwest Vegetable Production Guide for Commercial Growers.

Start with disease-free seeds, plants, and planting materials. It is important to start with disease-free seeds or planting materials (bulbs, tubers, transplants, sets, and so on) to keep from introducing serious diseases into the garden. In general, gardeners are encouraged not to save their own seed but to purchase seed from reputable seed dealers. Hot-water soaks and fungicide seed treatments may be used to control the seedborne disease organisms that cause seed rots and damping-off. Since temperature controls and timing must be precise, home gardeners usually buy seed that has already been treated with hot water by the seed processor. Hot-water treatment of cabbage, broccoli, Brussels sprout, and cauliflower seed is particularly important.

Many vegetable seeds are coated with a fungicide at the time of purchase, as evidenced by the coloured coating on the seed. Untreated seed can be treated by the home gardener by placing a small quantity (the size of one or two match heads) of captan or thiram in the seed packet and shaking the packet for a minute or two to thoroughly cover the seed. Excess protectant may be sifted out before planting.

Do not plant diseased material (for example, transplants, sets, bulbs, or tubers). All planting material should be healthy and free of yellowing and brown or black spots, and should not be stunted or show poor development. Only certified, disease-free potato tubers or sweet potato slips should be used. Examine transplants thoroughly for signs of leaf or stem disease. It never pays to buy and plant diseased transplants, no matter what the price. Damping-off diseases affecting home-grown vegetable transplants can be controlled by the use of diseasefree soil and fungicides.

Cultural Practices

Cultural practices such as controlling weeds, planting at the right time, planting at the proper depth and spacing, employing cages or stakes, and watering when the temperature is rising to promote a rapid drying of the foliage can help control many foliar diseases. Mulches help to control fruit rots and aid in control of blossom-end rot of tomato, pepper, and vine crops by maintaining a uniform supply of moisture in the soil. Weed control in and around the garden reduces the risk of attacks by viruses that overwinter in these plants. Do not work or harvest wet plants, since this can spread many diseases, particularly those caused by bacterial and fungal pathogens.

Disease Control During the Growing Season

Many diseases, such as the early blight disease of potato and tomato, occur each year despite all preplanting precautions. For such diseases, applications of fungicides and bactericides to the growing plants may be needed. The best way to apply these materials is as sprays before the disease occurs. Often a wetting agent such as liquid detergent or soap (1/2 teaspoon in 1 gallon) is added to obtain more thorough wetting and coverage of the foliage. Apply sprays to the point of run-off on a 7- to 10-day schedule. This maintains a fresh or effective covering of fungicide and protects the new growth. Fungicides and bactericides currently recommended for use on vegetables is given in the Illinois Pest Control Handbook which is updated each year.

Principal Diseases of Vegetables and Control Measures

Asparagus

Rust: Grow resistant varieties Start spray applications after harvest and continue until mid-August on a 10-day schedule with an approved fungicide.

Crown and root rots: Maintain proper fertility and good soil drainage. Avoid over-cutting.

Beans

Seed decay and damping-off: Use captan- or thiram-treated seed. Plant in a warm, moist, well-prepared seedbed.

Bacterial blights: Do not save seed from infected plants. Purchase seed from a reputable seed dealer. Spray a copper-based bactericide such as copper sulfate at the first sign of disease, and continue on a 7- day scheduled until harvest. Do not work among wet plants.

Rust. Grow rust-resistant varieties: Apply fungicides at the first sign of disease and continue on a 7-day schedule.

Mosaics. Grow resistant varieties..

Beets

Seed decay and damping-off: Plant seed treated with thiram or captan.

Leaf spots: Spray with an approved fungicide on a 7-day schedule.

Carrots

Seed rot and damping-off: Plant seed treated with captan or thiram.

Leaf spots: Practice a 2- to 3-year crop rotation. Spray weekly with an approved fungicide, starting about mid-June.

Aster yellows: Destroy infected plants as soon as they appear. Apply insecticide sprays for leafhopper control. Spray before removing infected plants.

Cole Crops: (cabbage, broccoli, Brussels sprout, cauliflower, Chinese cabbage, kohlrabi, mustard greens, radish, rutabaga, and turnip).

Seed decay, damping-off, black rot, and blackleg: Plant hot-water-treated seed which has also been treated with captan or thiram. Several black-rot-resistant cabbage varieties are now available.

Leaf spots and blights: Practice 2- to 3-year crop rotation for broccoli, cabbage, cauliflower, and Brussels sprout. Apply an approved fungicide on a 7-day schedule.

Clubroot: Purchase disease-free transplants. PCNB (Terraclor) may be added to the transplant water.

Fusarium yellows: Grow only varieties that are resistant to yellows.

Cucurbits (Cucumber, gourds, muskmelon, pumpkin, squash, and watermelon).

Seed rot and damping-off: Plant seed treated with captan. Plant in a warm, well-prepared seed bed.

Angular leafspot: Practice a 2- to 3-year crop rotation. Spray with a fixed copper bactericide, such as COCS. Remove diseased plant material. Plant resistant varieties whenever possible.

Anthracnose, Alternaria leaf blight, downy mildew, blossom blights, fruit rots and spots, and gummy stem blight or black rot. Practice 2- to 3-year crop rotation. Apply an approved fungicide at weekly intervals, starting when the plants are in the two-leaf stage. Adequate nitrogen fertility is essential for the successful control of Alternaria leaf blight. Plant resistant varieties when available.

Bacterial wilt: Spray with an insecticide to control cucumber beetles as soon as the seedlings "crack" the soil before they emerge. Continue weekly spraying until the plants are in bloom. Spray in the evening to avoid killing bees.

Fusarium wilt: Grow only cantaloupe and watermelon varieties that are resistant to Fusarium wilt.

Powdery mildew; Spray with an approved fungicide at the first sign of disease, and weekly thereafter.Grow resistant varieties.

Mosaic and other virus diseases: Grow resistant varieties where possible. Only cucumber-mosaicresistant cucumbers should be grown.

Eggplant

Seed rot, damping-off, and Phomopsis blight. Plant seed treated with captan or thiram.

Phomopsis blight and other fruit rots. Practice a 2- to 3-year crop rotation. Spray weekly with an approved fungicide when fruits are half size, or when disease first appears.

Verticillium wilt. Mulching with black plastic may help reduce losses.

Onion

Smut, seed rot, and damping-off. Plant seed treated with thiram or captan.

Leaf diseases: Practice a 2- to 3-year crop rotation. Spray weekly with an approved fungicide. Carefully follow label directions as regards the days between the last spray and harvest.

Bulb rots: Control leaf diseases. Also follow these recommendations: avoid storing improperly cured or injured bulbs; let all onion tops fall over (ripen) naturally; store under cold dry conditions; and do not allow cured bulbs to be exposed to sunscald, water, or high humidity.

Fusarium basal rot. Grow resistant varieties.

Pea

Seed rot and damping-off. Plant seed treated with captan or thiram.

Root rot: Plant early in well-drained soils. Use a fungicide seed treatment.

Fusarium wilt: Plant only varieties that are resistant to wilt.

Peppers

Seed rot and damping-off. Plant seed treated with captan or thiram.

Bacterial spot. Use the hot-water seed treatment, or dip seed in a solution of 1 part household bleach to 3 parts water and rinse with water. Follow with a fungicide seed treatment. Apply weekly sprays of a fixed copper fungicide such as COCS, through the bloom stage. Use crop rotation outside the tomato family

Tobacco mosaic. Plant resistant varieties.

Potato

Viruses, late blight, Verticillium wilt, and ring rot. Plant only "certified" disease-free seed.

Seedpiece rots. Treat cut seedpieces with captan as a dust or dip. Plant in warm soil (over 50°F) after the cut surfaces have corked over (suberized).

Early blight and late blight. Practice a 2- to 3-year crop rotation. Destory or remove cull tubers. Spray at weekly intervals, starting when the plants are 4 to 6 inches high with an approved fungicide. Plant late blight-resistant varieties.

Verticillium wilt and root-knot nematodes. Plant resistant varieties.

Scab. Plant resistant varieties, or adjust the soil pH to 5.2 to 5.5. Use fungicide seedpiece treatments.

Storage rots. Store only disease-free, unbruised tubers. Allow tubers to cure at 60°F for 5 to 7 days before reducing temperature to 36° to 40°F (if possible).

Sweet corn

Seed rot and damping-off. Plant seed treated with captan or thiram.

Smut. Plant resistant varieties. Remove smut galls from the garden before they break. Do not compost smutted corn or infected corn residues.

Stewart's bacterial wilt. Plant resistant varieties. Control flea beetles with a recommended insecticide. Early season control is particularly important.

Foliage blights and rust. Plant varieties resistant to northern and southern corn leaf blights and rusts. Apply an approved fungicide. However, the use of fungicides is rarely justified. Diseased plant material should be removed from the garden. Blighted leaf tissue is safe to compost.

Maize dwarf mosaic. Plant as far from Johnsongrass as possible. Control wild Johnsongrass plants. Plant tolerant varieties.

Spinach

Seed rot and damping-off. Plant seed treated with captan or thiram.

Cucumber mosaic virus (blight). Plant resistant varieties.

Downy mildew or blue mold. Plant resistant varieties.

Sweet Potato

Black rot, scurf, foot rot, wilt, and soil rot. Buy certified, disease-free slips. Plant resistant varieties. Use a 3- to 4-year rotation.

Storage diseases. Do not expose roots to temperatures below 55°F. Cure the roots immediately after digging at 85°F and high humidity for 10 to 14 days. Store at 55°F. Never store in airtight bags or containers. Apply an approved fungicide as a post-harvest dip. Follow the manufacturer's directions carefully.

Tomato

Seed rot, damping-off, bacterial spot, bacterial canker, and tobacco mosaic virus. Plant seed which has been treated with hot water or acid and then treated with captan or thiram. Purchase seed or transplants only from reputable dealers. Soaking seed in copper sulfate solution (2 ounces per gallon) or dipping in bleach solution as described under peppers will help prevent bacterial spot and canker if seed are saved.

Bacterial spot: Plant certified, disease-free seed or transplants. Practice a 2-to 3-year crop rotation. Spray plants weekly with a fixed copper fungicide, beginning with the first sign of disease, through the bloom stage.

Early blight, Septoria leaf blight, anthracnose, buckeye rot, and soil rots. Spray weekly with an approved fungicide. Practice 2- to 3-year crop rotation. Remove or destroy infected plant parts. Carefully follow label directions as regards the days between the last spray and harvest. Staking or caging plants and providing good air circulation will help reduce foliar diseases. Staking, caging or mulching will also help prevent fruit rots, such as anthracnose, buckeye rot, and soil rot.

Blossom-end rot: Maintain uniform soil moisture by mulching the plants and irrigating when needed. Avoid heavy pruning, excessive nitrogen applications, or damage to the roots by cultivation or hoeing.

Verticillium and Fusarium wilts: Grow only VF- or VFN-resistant varieties. There are many varieties available.

Root-knot nematodes: Grow VFN-resistant varieties.

Viruses: Avoid contact between potatoes and tomatoes to prevent "double infections" of potato and tomato viruses. Wash hands thoroughly with soap and hot water before handling your plants. Do not use tobacco when working with tomatoes.

Disease Control in the Home Vegetable Garden

Plant diseases can be a significant problem in home gardens. Most vegetables are susceptible to a number of diseases. Wilts, leaf spots, blights and fruit rots are just a few of the disease problems that plague vegetables gardens every year.

Plant diseases are caused by four primary types of organisms: fungi, bacteria, nematodes and viruses. These organisms are often referred to as *plant parasites* or *pathogens*.

Plant diseases caused by fungi and bacteria are most common when rain showers and/or heavy dews are frequent and temperatures are warm. During these times, scout the garden regularly for disease. Viral diseases can occur at any time, and nematode damage can be more evident if conditions are dry. Home gardeners can reduce the occurrence of many diseases with sound cultural practices.

Site Selection

Although few people have an ideal garden spot, site selection is important for a successful garden.

A home garden site should be well drained. Avoid wet and poorly drained soils. Excessive soil moisture will favour root and crown diseases such as damping-off, crown and root rots caused by fungal soil-borne pathogens.

Most vegetables require full sunlight for maximum yield. Full sun speeds drying of the foliage, which reduces the incidence of most foliar diseases.

Crop Rotation

Crop rotation is very important in reducing losses to vegetable diseases. Continuous plantings within the same plant family of vegetables provide opportunities for pathogen buildup. Do not plant the same family of vegetable in the same areas year after year. Grow the same plants or closely related plants in the same soil only once every 3 to 5 years. This practice starves out most pathogens that cause stem and leaf diseases.

Unfortunately, crop rotation does not work against soil-borne problems such as root and crown diseases caused by the fungi *Phytophthora, Rhizoctonia, Pythium* and *Sclerotium,* and vascular wilts caused by the bacteria *Pseudomonas.* These organisms are long-lived in the soil and affect many plant families. You can get only a limited benefit against *Fusarium* wilt.

Disease Free Seed and Transplants

Many plant diseases can be seed-borne. Do not save seeds from year to year. This is important to prevent a number of diseases, including halo blight, common blight, anthracnose of snap beans, and bacterial spot of tomatoes. Commercially available seed is often produced in the arid western United States where seed-borne pathogens do not occur and/or cannot survive. Seed of some vegetable varieties, however, are not produced in the west, and local seed must be used. Purchase seed from a reputable dealer because you cannot distinguish healthy seed from diseased seed. Seed are protected by fungicides often applied to the seed by the company. Seed can also be treated by the home gardener. These treatments give young seedlings some protection from soil-borne pathogens as they germinate and emerge. They do not, however, control pathogens that may attack the plant after the seedling stage. Some pathogens are carried on the surface of seed and can be controlled with fungicide treatment. Fungicides are not effective on pathogens carried beneath the seed coat.

Gardeners starting a crop from transplants should examine transplants carefully before purchasing. Select healthy transplants that appear disease and insect free.

Use Resistant Varieties

Using resistant varieties is the most efficient way of controlling vegetable diseases. Make an effort to buy resistant varieties when they are available. Seed catalogs generally list resistant traits of different vegetable varieties.

For example, disease resistant tomatoes are designated "VFN." These letter indicate resistance to Verticillium wilt (not common in Georgia), *Fusarium* wilt and nematodes. In tomatoes, there is also resistance to *Stemphyllium* (gray leaf spot) = S, Tobacco Mosaic Virus = TMV, *Alternaria alternata* stem canker = A or ASC, *Septoria* leaf spot = L and Tomato Spotted Wilt = TSW.

Other vegetables such as squash, pepper, beans and peas also have disease and nematode resistance. Note that plant disease and nematode resistance are relative terms, as resistant varieties have differing levels of resistance. Resistance does not always mean the plants are immune to these problems. Some seed catalogs may indicate tolerance to a certain pest. Tolerance usually indicates that a particular variety of vegetable may produce fruit of acceptable yield and quality while exhibiting moderate to high levels of disease.

Planting Date Management

Planting dates can be an effective tool for disease management. Follow the recommended planting dates for the particular vegetable grown. For example, okra should be planted when soil temperatures are warm for good germination and growth. Planting this crop when soil temperatures are cool can cause increased incidence of soil-borne diseases such as soreshin and damping-off. Using transplants may help avoid early disease problems so long as the transplants are from a reputable source and are disease free. Plant crops such as corn and beans earlier, because their seeds germinate in cooler soils. It is also advantageous to plant these particular crops early to escape severe virus infections. This is particularly true with squash. Aphids, which transmit many virus diseases, are at lower population levels early in the season. High aphid populations late in the season can lead to more virus problems.

Trap Crops

Using trap crops may sometimes help manage virus diseases and aphid populations. A few rows of a trap crop around the vegetable garden (such as rye or corn) will cause aphids to feed there first, possibly loosening the virus they may carry. Trap crops can limit aphid damage and help reduce the incidence of virus diseases, which are sometimes severe on beans, cucurbits and solanaceous crops.

Proper Spacing and Trellising

Proper spacing and trellising can reduce the occurrence of vegetable diseases. Space plants properly to allow growth and air circulation. Humid and wet conditions occur if plants are crowded and unable to dry quickly. Extended periods of wet foliage are ideal for development of fungal and bacterial diseases. Crowded plants take longer to dry and provide favourable disease conditions. Crop protectants also cannot penetrate dense foliage, thus limiting their effectiveness.

Staking or trellising tomatoes, pole of half runner beans, and cucumbers will prevent soil contact with the foliage and fruit. Air circulation will be better if these plants are trellised, promoting foliage drying and reducing disease severity. Limiting soil-plant contact reduces diseases such as fruit rots. Crop protectants also can be more effectively applied to trellised plants.

Proper Watering Practices

Avoid excessive soil moisture. Over-watering enhances seed decay, damping-off, and root and crown rot diseases. *Choanephora* fruit rot, which causes the whisker-like blossom end rot, can be increased by over-watering.

If possible, irrigate by running water between the rows. Avoid splashing plants with soil, since any potential pathogens are soil-borne. Drip irrigation is also a good method of watering without wetting foliage.

Drip irrigation uses either a hose or tape that allows water to be applied slowly over time at the base of the plant or just beneath the soil surface. Garden drip irrigation kits are readily available.

Avoid wetting the foliage when watering. Wet foliage is favourable for development of most foliar fungal and bacterial diseases. If wetting the foliage is unavoidable, irrigate in the morning so foliage can completely dry before evening.

Maintain uniform soil moisture. This can reduce problems such as blossom end rot (calcium deficiency) of tomatoes and peppers.

Don't work in the garden when plants and soil are wet. Bacterial and fungal diseases spread easily from one plant to another by hands and clothing when above-ground plant parts are wet.

Use a Mulch Layer

Use a mulch layer of straw, bark, shredded paper or plastic to prevent soil from splashing onto plants and to prevent fruit from touching the bare ground. This will help prevent rots on mature fruit such as strawberries, tomatoes, squash, cucumbers and melons. Mulches also are a sound cultural practice to help conserve soil moisture and reduce weed infestations.

Proper Fertilization

Adequate and proper fertilization helps prevent vegetable diseases. Test soil three to six months before the growing season, and follow the recommendations to supply appropriate nutrient requirements and adjust soil pH. Avoid excessive amounts of nitrate forms of nitrogen, which encourage root rot diseases. Use ammonium forms instead (no more than 50 percent). Be sure the soil pH level is in the proper range for a particular crop. Proper pH prevents blossom end rot and encourages healthy growth of tomatoes and peppers.

Weed Free Garden

Weeds in the same families as some of the vegetable crops can be another source of diseases and insects. For example, some weeds can serve as virus reservoirs for several insect-transmitted viruses that can infect homegrown vegetables. Good weed control will increase air movement in the garden and decrease conditions — such as excessive moisture — that favour disease development.

Avoid Tobacco when Working in the Garden

If you use tobacco, wash your hands thoroughly before handling plants. This practice will prevent the spread of tobacco mosaic virus, which can infect many different kinds of vegetables, particularly solanaceous crops such as tomatoes and peppers.

Nematode Control

Nematode control is important in growing vegetable crops. Nematodes are microscopic soil-inhabiting roundworms that injure vegetables and other plants by feeding on their root systems, causing decay and/or galling. You can sample soil for nematodes by submitting a sample through your county extension office to the Extension Nematology Laboratory.

The most damaging nematode in Georgia is the root-knot nematode, named for the knots it produces on infested roots. Soil solarization can be used to reduce the number of nematodes present. This involves tilling the garden and then covering the area with a clear plastic tarp for six to eight weeks.

Solar radiation is trapped in the soil and can raise soil temperature high enough to kill most nematodes in the upper 3 to 4 inches of soil. In Georgia, the best time to solarize soil is June through August, when temperatures are at their hottest. Soil solarization requires intense solar heat to be effective. Repeated roto-tilling can also desiccate and kill many nematodes, although repeated tilling may decrease overall soil organic matter.

Sanitation

Sanitation will help reduce overwintering of disease-causing organisms. After harvest, remove and destroy plant material. Plow the soil to help break down small roots and debris that may harbor nematodes, fungi and bacteria. Plant refuse may also be plowed under. Remove diseased plants, plant residue and weeds in and around the vegetable garden to reduce the occurrence of some diseases.

Pesticide Use

Use of pesticides should be the last defence available to home gardeners once all other disease-control options have been exhausted. Home gardeners have few fungicides or bactericides available compared to commercial growers.

Chlorothalonil (Bravo® or Daconil®), maneb, mancozeb, Terraclor® (PCNB), sulfur and copper products can be used on certain crops. These products provide disease suppression across a range of foliar diseases and are more effective if they are applied in a preventive manner at the very onset or before a disease outbreak occurs. Chlorothalonil, maneb and mancozeb generally provide broad-spectrum disease suppression of most fungi. Terraclor is used as a transplant soil drench to suppress

Southern blight and Rhizoctonia damping-off. Copper products suppress fungi but primarily reduce losses to bacterial pathogens. Sulfur is suppressive to fungi and is especially effective against powdery mildews. As with any pesticide, read the label and follow the recommended precautions to ensure the chemical is applied in a safe, effective manner.

Contact your county extension office for specific information about pesticides.

Most importantly, use proper cultural practices to promote and maintain healthy plants. Healthy plants do not get diseases as readily as weak ones. Healthy plants are the best control against plant diseases.

Natural Pest and Disease Control in Your Vegetable Garden 0

Controlling pests organically means much more than simply changing the types of sprays and dusts you use. Organic gardeners aim to create a balanced system where pests are regulated naturally and where there is seldom the need to use even the safest organic sprays and dusts.

One of the principles at the heart of organic gardening is to completely eliminate the need to use sprays and dusts, even organically acceptable ones. While this might seem foolish — or even impossible — at first, by removing the need to continually spray and dust your garden, producing your own food becomes more enjoyable, safer and will require less work. This means more time spent admiring and eating the fruits of your labour, and less time spent preparing concoctions and potions!

So where do you start on the road to organic, pest-free self-sufficiency? Surprisingly, some of the best ways to contol pests and diseases don't at first glance appear to be controls at all. By looking at her vegetable garden as a mini ecosystem which needs to be kept in balance, and thus in which no pests or disease can get out of control, the organic gardener is able to concentrate her efforts on:

- cultural techniques which will keep her vegetable plants in the best of health
- Natural Pest and Disease Control in your Vegetable Garden 0

- developing rich, biologically active soil
- Natural Pest and Disease Control in your Vegetable Garden 0
- encouraging a diverse population of predator species to keep insect pests in check.

But the place to start, and the foundation of your future organic vegetable growing success, starts with the plants themselves.

Buying Healthy Plants

Starting out with the healthiest plants you can find is a good way to ensure a healthy and productive vegetable garden. Use these tips to choose the best plants you can find:

Buy from nurseries or market stalls where the plants are well cared for. Plants which are allowed to stand in the sun on a hot pavement or which have been allowed to wilt are not a good choice, even if they're cheap.

Look at the entire selection of plants on offer. If some of the plants appear to be in poor health, it may be that the healthier looking ones have arrived more recently from the supplier and have yet to succumb to the seller's mistreatment. They might look healthy today, but could very well be diseased tomorrow.

Carefully look at the plant's colour. Plants which appear slightly pale might just need to be fertilized, and this is easily corrected. But plants which have distinct yellow streaks or brown spots are diseased. Do not buy them, or any other plants in the selection.

Look for roots, and look at the roots if possible. Plants which have masses of roots growing through their container's drainage holes have been left in too small a container for too long, and this could result in severe transplant shock if much of the root mass needs to be boken off in order to remove the plant from its container before transplanting.

If you are able to (and if you are buying large quantities of plants, insist that you be allowed to) gently shake a plant loose from its container so that you can exmine the roots. The roots should be plentiful, but not wrapped in a tight spiral or almost filling the container.

Cultural Techiques

Water early: If you irrigate your garden with a sprinkler from overhead, it's best to water early in the day so plants can dry off before night falls. Foliage that stays wet for long periods of time is susceptible to leaf diseases, fungi that grow on leaves, tender stems, and flower buds. This tends to be a problem when plants stay wet throughout the night: Fungi spread quickly during the cool, moist evening hours. The fungi will cause the plant to be weakened, flowers will fall off, and fruit will begin to spot and become soft

Crop rotation: Do not grow the same plant family in the same spot year after year. Repetition of the same crop gives diseases a chance to build up strength. Design your garden so that each family of vegetables — cabbage family, cucumber family, and tomato/pepper family — can be moved to another block of your garden on a three-year rotation.

Avoid monocultures: Planting a wide variety of plants in your garden is another way to foil pests and diseases by encouraging diversity. Conventional lawns are an excellent example of a monoculture. Planting a mixture of lawn grasses will help prevent diseases from sweeping through your lawn, because some grasses will be susceptible and others won't.

Companion planting is another way to use diversity – and avoid monocultures – to foil pests. Mixing marigolds and herbs in and between your vegetable plants, as opposed to planting solid blocks or rows of a single vegetable crop, is very effective, because many pests locate crops by smell and these strong-smelling plants can "hide" a crop by masking the odour which attracts insects.

Developing Rich Soil

Soil that is fertile, well-drained and teeming with communities of a diverse range of microorganisms will greatly increase your vegetable garden's health and productivity, since many pests and diseases spend part or all of their lives below-ground. Having a diversity of microbes in the soil will help keep them in check.

Healthy soil is an intricate mix of tiny rock particles, organic matter, water, air, fungi, microorganisms and other animals. The more organic matter you add to your soil, the more biologically active your soil will become, and the more active it is, the greater the competition between benign microorganisms and those intent on doing harm to your plants. Over time, adding organic material improves the soil's structure, which in turn improves its ability to absorb and release water, thereby maintaining ideal moisture levels.

Although rich, well-fed soil is all most vegetable plants need to keep them growing vigorously and producing well, fertilizing them at the correct time is also beneficial, and will help boost their natural resistance to pests and disease. As a general rule, annual vegetables should be well-fed when they are young, and then given "booster" feeding when they start producing. Perrenial vegetables should be fertilised during the first third of their active growth in spring.

You can also fertilise them in summer, but not in autumn, as late fertilisation can stress them by encouraging tender new growth which may be winter-killed. When you apply fertiliser, always keep in mind that more is never better. Over-fertilisation leads to rank, spindly growth which always worsens problems with aphids and some other pests.

Encouraging Predators

Encouraging a diverse community of insect predators is one of the best ways to manage pests. Beneficials, the term used for insect predators and parasites, play a vital role in the complex community which naturally exists in your garden by eating or parasitizing harmful pests. The first and most important rule for the gardener who wants to encourage beneficial insects is to immediately stop using toxic sprays and dusts. Even organic pesticides kill indiscriminately, so use them only when absolutely necessary and then only on the plants being attacked. To attract beneficial insects into your garden, provide them with an attractive food supply. Since the adults of many beneficial insects only feed on pollen and nectar, the best way to attract them is to plant small-flowered plants such as dill, fennel, parlsey and mint.

You can also allow a few of your vegetables, especially members of the mustard family such as radishes, and broccoli to go to flower. There are also many ornamental annuals which can be used: simply use them as colourful borders or edging plants around your vegetable beds. Once the adults of beneficial species have arrived in your garden and had a good meal, the females will search for good places to lay eggs which will hatch into predatory larvae. Beneficial ground-dwelling insects such as beetles find refuge in the soil, under stones and in thick mulches.

It is important to minimise dust and to provide a source of water in hot, dry areas to protect beneficial insects, which are easily killed by dehydration. An old birdbath filled with water and rocks (to provide safe landing places and prevent drownings) will be used by many beneficial insects.

Birds are amongst the most efficient predators of harmful insects, so planting shrubs and trees, and supplying them with a source of water to attract them to your garden – and encourage them to stay and eat their fill!- is a worthwhile investment.

And for consuming slugs, cutworms and many other ground-dwelling pests, nothing beats frogs and toads. Be sure to provide shelter and water for them, too.

Non-toxic Control of Insects

There are many different non-toxic control methods we can use against insects in the garden. Some are commercial sprays and powders, others can be made at home from household ingredients. Use non-toxic insecticides only when absolutely necessary The following list of non-toxic insecticides should only be used whco absolutely necessary. It's important to allow the garden to develop it's own micro-divcrsily and ccology. *Bacillus thuringiensis* of the microorganisms that make insects ill, Bacillus thuringiensis is by far the best known. Though safe to humans, it has a devastating effect on many pests. BT, as it is often called, appears on the market in the form of a soluble powder. The powder is suspended in water and sprayed on plant surfaces, where it is ingested by the pest. The BT spores then germinate into plants that occupy more and more of the victim's body. A toxic crystal is also produced and is thought to be partially responsible for the insecticidal effect. BT is useful on both vegetable and fruit crops, aiding in the control of caterpillars, earworms, peach tree borer, cabbage moth, cabbage looper, and gypsy moth caterpillar.

Pyrethrum A is a botanical insecticide derived from the flowers of a species of chrysanthemum. The material causes rapid paralysis of most insects, but the insects usually recover unless the pyrethrum is combined with a synergist or other poison. There are several pyrethrmun products available that are certified organic. Use against aphids, thrips, caterpillars, ants, flies, earwigs, cabbage moth. As pyrethrum also affects beneficial insects it should only be used as a last resort.

Wettable sulphur is an effective fungicide and miticide. The sulphur interferes with important processes in fungal spores and is noted for its ability to suppress mite populations.

Chapter 9

Emerging Virus Diseases of Greenhouse Vegetable Crops

There have been several new viral diseases reported on greenhouse vegetable crops worldwide, and some of them are of great concern to greenhouse vegetable production in British Columbia. Information given below focuses only on a few potential new viral diseases that have been reported on greenhouse vegetable crops in North and Central America, and, therefore, these viruses are very likely to be introduced into Canada.

The purpose of this factsheet is to bring awareness to the greenhouse vegetable growers and to engage them in active surveillance and biosecurity measures to prevent the introduction, establishment and spread of such damaging viral diseases.

Early detection of viral diseases is critical to prevent or minimize the spread of the virus and potential crop losses in greenhouses. If a new viral-like symptom(s) or a potential virus-vector is suspected in a greenhouse, then, it is critical to confirm the presence of it by following a proper identification process through a recognized plant pest diagnostic laboratory.

Viruses are too small to be seen and identified with a light microscope. Therefore, specialized techniques such as ELISA (enzyme-linked-immunosorbent-assay) and RT-PCR (reverse transcriptase polymerase chain reaction) are required in order to confirm the virus.

Once the identity of the virus is confirmed, specific control strategies must be implemented immediately.

Columbian Datura Virus (Potyvirus)

Virus Description and Distribution

Columbian datura virus (CDV), a member of the family Potyvirus, is known to infect plants particularly in the family *Solanaceae*, including tomato and tobacco, and many ornamentals (*Brugmansia, Juanulloa, Petunia, Pepino* etc.). Initially, CDV was detected in *Datura* species imported from Columbia to the United States in 1968. Since then, the virus has been reported from Australia, Germany, Hungary, Japan, Netherlands, Poland and the United States. In the United States, CDV was reported from several States in 2003 and 2004 on ornamental plants (*Spiranthes* orchids, *Brugmansia* spp. etc.). In Canada, CDV was first detected in 2007 on *Brugmansia* spp. In British Columbia. CDV is listed on the USA Regulated Plant Pest list but not on the Canadian Regulated Plant Pest List. It appears that CDV is more widespread than originally thought. Economically important Solanaceous crops, including field-and greenhouse-grown tomato and field potato can be at risk. It is important that growers pay close attention to the disease and engage in periodic monitoring for symptoms on greenhouse tomato.

Tomato Infectious Chlorosis Virus (Closterovirus)

Virus Description and Distribution

Tomato infectious chlorosis virus (TICV) is a member of genus Crinivirus (family Closterovirus). It was first reported in 1993 on field tomatoes and then in greenhouse tomatoes in California and North Carolina, USA. Subsequently, the virus was detected on greenhouse tomatoes in Europe (Italy, Spain, Greece and France) and Asia (Indonesia, Japan and Taiwan). It has been shown to have a moderately wide host range (26 plant species in 8 plant families), including vegetables such as tomato, potato, lettuce and artichoke, ornamentals such as petunia, and several weeds.

Tomato Chlorosis Virus (Closterovirus)

Virus Description and Distribution

Tomato chlorosis virus (ToCV) is also a member of genus Crinivirus (family Closterovirus) which is closely related to

Tomato infectious chlorosis virus (TICV). ToCV causes symptoms on tomatoes that are very similar to the symptoms caused by TICV and, thus, cannot be distinguished. However, ToCV and TICV can be differentiated based on the symptoms expressed on the indicator plants, *Nicotiana benthamiana* and *N. clevelandii*.

The major difference between ToCV and TICV is that ToCV is transmitted by the greenhouse whitefly (*Trialeurodes vaporariorum*), the banded-wing whitefly (*T. abutilonea*) and silverleaf whiteflies (*Bemisia* spp.), biotypes A (*B. tabaci*) and B (*B. argentifolii*), while TICV is transmitted solely by the greenhouse whitefly.

Like TICV, ToCV is also transmitted in a semi-persistent manner by whiteflies and its movement in host plants is restricted to phloem tissue. ToCV has been shown to have a moderately wide host range (24 plant species in 7 plant families), including tomato and sweet pepper, as well as some weeds.

Although ToCV first appeared in 1989 in greenhouse-grown tomatoes in Florida, USA, and, later, in Colorado and Louisiana, it wasn't confirmed as ToCV until 1996. Subsequently, ToCV has been reported on greenhouse-grown tomatoes in many countries of the Caribbean, East Asia, Southern Africa, Europe and the Mediterranean. In 2003, sweet pepper was identified as a natural host for ToCV in Spain.

Symptoms

Tomato plant infected with ToCV expresses symptoms similar to those expressed by TICV. Therefore, it is difficult to diagnose plants infected with these viruses based on the symptoms. It is also possible that a single tomato plant can also be co-infected with both ToCV and TICV. It is, therefore, critical to submit suspected plant samples to a recognized plant pest diagnostic laboratory for accurate identification.

Spread

TICV is transmitted in a semi-persistent manner by the greenhouse whitefly (*T. vaporariorum*), the banded-wing whitefly (*T. abutilonea*) and silverleaf whiteflies (*Bemisia* spp.); biotypes A (*B. tabaci*) and B (*B. argentifolii*). The movement of virus in the host-plant is restricted to phloem tissue. Studies have shown that persistence and efficacy of transmission of ToCV is variable depending on the type of virus-vector involved. Although all ToCV vectors are capable of transmission, the banded-wing whitefly and silverleaf whitefly-biotype B are highly efficient vectors of ToCV. The virus is not seed-borne or transmitted mechanically.

Tomato Yellow Leaf Curl Virus (Bigeminivirus)

Virus Description and Distribution

Tomato yellow leaf curl virus (TYLCV) is known to infect many vegetable crops including tomato, pepper and bean as well as many ornamental plants such as poinsettia. TYLCV can cause severe economic losses to tomato production, where yield losses of up to 100% in greenhouse tomato production have been reported in Southern Europe and the Middle East. TYLCV spreads systemically in the host plant and is limited to phloem tissue.

TYLCV has been a major threat to tomato production in Israel since 1930. It was introduced into the Dominican Republic, Cuba, and Jamaica in the early 1990s'. Since then, TYLCV has been reported from Africa, Asia, Australia, Caribbean, Europe and North and Central America. In North America, TYLCV is known to occur in Florida, Georgia, Louisiana, and has recently been detected in California and Arizona.

Symptoms

Visible symptoms can be seen on tomato plants 2-3 weeks after initial infection by TYLCV. Symptoms can vary slightly depending on the tomato cultivar and amount of virus infection. Generally, infected plants have smaller-than-normal leaves that are cupped upward, thick and rubbery with chlorotic margins. Young infected plants become severely stunted. Infected plants drop up to 90% of their flowers resulting in major yield losses.

Spread

TYLCV is transmitted by adult silverleaf whiteflies, *Bemisia argentifolii*, previously known as *Bemisia tabaci* biotype B,. The silverleaf whitefly is a different species than the common greenhouse whitefly, which does not transmit TYLCV. Silverleaf whiteflies are rarely seen in British Columbia. They are small, phloem-feeding insects which pick up the virus while feeding on infected host plants. Since TYLCV has a broad host range, it can be spread by silverleaf whiteflies that feed on other infected host plants, including cultivated and wild plants. Not all host plants infected with TYLCV show visible symptoms, however the virus can be acquired by silverleaf whiteflies from symptomless infected plants and transmitted to tomato. The virus can also be transmitted through cuttings taken from infected plants or it can be introduced into a greenhouse with virus-infected transplants.

TYLCV can not be transmitted mechanically or by workers and is not transmitted by seeds. Even if TYLCV shows up in British Columbia, it may not become established due to the unlikely survival of silverleaf whiteflies in winter months. On the other hand, there is evidence to suggest that TYLCV can be persistent and can overwinter in infected plants and plant materials.

General Virus Control Measures

Virus diseases are very difficult to control once they become established. The possible presence of a potential virus on other host plants (including weeds) in the absence of a host crop, and the ability of virus-vectors (the greenhouse whitefly, the banded-whitefly and silverleaf whiteflies) to feed and reproduce on crop plant make it difficult to control the disease successfully. It is highly recommended that a collective management approach, i.e. a combining of cultural practices, biosecurity measures, greenhouse sanitation and disinfection practices, and insect-vector control strategies, should be adopted to reduce the impact of virus diseases on a crop.

- Use virus-free transplants that come from certified transplant nurseries
- Use tomato cultivars, if any, that are resistant to TICV, ToCV, TYLCV and CDV
- Monitor for whitefly populations throughout the tomato growing season (spring to fall) and, if present, apply appropriate insecticides that are registered in British Columbia for greenhouse tomato. Protection of young

tomato plants from whitefly-feeding is essential to reduce potential yield losses.

- If whiteflies are detected in British Columbia greenhouses, use reflective or coloured (yellow) mulches that may reduce the whitefly populations feeding on tomato leaves.
- Use trap plants, preferably cucurbits that are preferred by whiteflies, to reduce infection rate on tomato.
- Maintain periodic scouting for virus symptoms, particularly on young tomato plants, and remove and deep-bury or incinerate the infected or suspected tomato plants.
- Maintain healthy growth of tomato plants to minimize virus damage.
- Since viruses have a broad host range, including many weeds that are symptomless, maintain a rigorous weed control program in and around the greenhouse during the growing season and winter months.
- TICV, ToCV, TYLCV and CDV are not yet detected on greenhouse tomato in British Columbia and it is very difficult to eradicate once established. Therefore, preventing the introduction and establishment of these viruses is very important. Submit any suspected virus-infected plant samples or plants showing virus-like symptoms and any suspected virus-vectors (e.g. whiteflies) to the BCMAL Plant Diagnostic Laboratory or to a recognized plant pest diagnostic laboratory for proper diagnosis and confirmation.
- To prevent the introduction and spread of potential diseases of greenhouse vegetable crops, please refer the factsheets on "On-Farm and Greenhouse Sanitation and Disinfection Practices" and "Biosecurity Guidelines".

Rose Diseases: Bacterial and Viral

In any Integrated Pest Management program, proper identification of the problem is Step 1. Rose diseases are no different. They generally fall into one of three categories-bacterial, viral or fungal.

Bacterial Diseases

Infection can occur in most any garden. The pathogen enters through wounds to the plant caused by improper and unsanitary pruning, grafting, mechanical injury during cultivation, or failure to prune away broken stems and roots during transplanting. Bacteria can survive in the soil for at least two years. Besides spreading via infected pruning equipment, insects and water are also known vectors.

Crown Gall-Unlike the galls formed by stinging insects which are normally harmless to the tree or plant, Crown Gall is caused by a bacteria. These galls begin with tumour-like cell growth at or just below the soil's surface, near the base of the plant and commonly on bud unions. Galls usually begin as green, pliable tissue; then develop into dark, crusty growths. Following gall formation, shrub growth is stunted, foliage is sparse, and bloom production is reduced.

Root Gall-Like Crown Gall above, but damage is confined to roots. Commonly caused by improper root pruning during transplant. Stem Gall-Like Crown Gall above, but damage is confined to stems and canes above the soil's surface. Control of bacterial diseases is achieved by checking existing and newly introduced plants for contamination. Use caution when cultivating fertilizers, organic materials or mulches around the rose's root zone.

If galls are discovered, the plants should be removed as soon as possible. Replace contaminated soil with well-draining soil high in organic matter and beneficial microbial activity. When planting bare-root roses, prune broken or damaged roots and stems with disinfected pruning equipment. Dip pruning tools with a 10% liquid chlorine bleach and water solution.

Disinfect tools between each plant when pruning non-infected plants; between *each cut* when pruning infected plants. (Note: This may discolor metal parts. After disinfecting, scrub discoloured areas with steel wool, sharpen cutting edges, and oil metal surfaces to restore them.) Some rosarians also report success by pruning away galls and spraying the infected area with an anti-bacterial solution like Agri-strep or a copper compound like Kocide.

Raspberry Diseases

The most common fungal diseases include anthracnose, cane blight, spur blight, and botrytis fruit rot commonly called 'gray mold.' These diseases can cause defoliation, wilting of lateral shoots, death of fruiting canes, and a reduction in winter hardiness. Other diseases such as phytophthora root rot, powdery mildew, rust fungi, fireblight, leaf spot, and verticillium wilt can also infect raspberry stands, but are not usually considered as important or life threatening. The severity of each disease fluctuates yearly, depending on annual weather conditions and cultural practices.

To maximize raspberry production, plant disease-resistant or fall-bearing cultivars such as Heritage and Autumn Bliss on an appropriate site, and follow a thorough disease-management program. Frequent scouting and knowledge of the common diseases will help identify the correct disease organism and appropriate control measures. Many of the common diseases can be minimized or eliminated with cultural practices and sanitation, since most disease-causing organisms survive winter on infected plant material. However, other diseases will require fungicide applications.

To promote healthy raspberry stands and reduce plant disease, follow these cultural practices.

- Start with disease-free and virus-tested raspberry plants from a reputable nursery. (For more information on cultivar selection, see extension publication FS-1108-GO, Raspberries for the Home Garden.)
- Plant raspberries in well-drained soil or raised beds.
- Avoid overhead irrigation, over-fertilizing, and wounding plants.
- Control weeds within and around the raspberry stand.
- Prune and thin canes annually during dry weather to maintain optimum spacing and row widths.
- Destroy abandoned and wild brambles surrounding the raspberry stand, since they harbour disease-causing organisms.
- Remove newly infected canes and destroy.
- Remove old fruiting canes (after harverst) and destroy.

Fungal and Bacterial Diseases

Anthracnose is frequently observed on the canes of black raspberry and susceptible cultivars of red raspberry. In late spring, red-purple oval lesions appear on the primocanes. The centre of these lesions later turns pale brown to ash gray, while the margins become raised and purple. Lesions usually girdle the cane causing it to die, dry, and crack. Infected primocanes that survive winter usually produce irregular, lateral branching and irregular fruit.

Surviving canes may also suffer tip dieback. Spores produced on infected canes during wet periods in the spring initiate new infections. Anthracnose also infects the leaves, flowers, and fruit. Small, purple-brown spots typically develop on infected leaves.

Cane blight is usually associated with wounded or damaged canes. Spores produced throughout the growing season on previously infected floricanes are rain splashed to healthy primocanes, where new infections begin. Primocanes that survive the winter rapidly wilt and die the following season. Visual symptoms are typically expressed the following spring as lesions on the floricanes.

These lesions appear black, brown, or gray and contain black pimple-like pycnidia. Lesions often extend through several internodes on one side of the cane. Infected canes tend to be brittle and often break near the lesion. Cane blight causes wilting, death of axillary buds and lateral branches, and death of the cane. Spur blight is most frequently observed as lesions surrounding the nodes on primocanes. Infection initially begins along the margins of the leaf and progresses inward, through the petiole, to the stem.

Infected primocane leaves develop a brown 'V' shaped lesion with yellow borders, and even tually fall from the plant. Floricane leaves are also susceptible, but rarely infected. Cane lesions appear purple under the heavy primocane wax, but turn brown when the wax is rubbed off.

In winter, lesions turn silver to gray and become difficult to identify. In the spring, spur blight causes buds near infected nodes to fail or remain inactive. Infected lateral shoots (spurs) may also show a reduced number of flowers. Spur blight infects only the outer portion of the stem (cortex), so healthy green tissue can be found beneath the lesion.

Gray mold, a common and serious disease of raspberries, causes blossom blight and fruit rot. Gray mold can also infect senescing leaves and canes. Infection typically begins on young flowers, but quickly spreads to fruit during cool, wet weather. Often immature berries, with latent infections, remain symptomless until the fruit matures. However, when environmental conditions are cool and wet, infected berries may become covered by a mass of gray or brown fungal material.

This mass of fungal material gives fruit a moldy appearance. Physical damage, bruising, or wounding causes an increase in the number of fruit infections and post harvest rot. Cool, wet weather prior to and during harvest also increases the chance for rot.

Phytophthora root rot is caused by a soilborne fungus that commonly infects the roots of red raspberries. This disease causes the leaf margin to appear yellow, wilted, or scorched. Infected canes may wilt and die rapidly or remain symptomless until the following year. Infected plants may contain dark water-soaked lesions near the base of the cane or reddish-brown root tissue. Phytophthora can persist in the soil for many years and is stimulated by wet, saturated soils.

Verticillium wilt is a soilborne fungus that infects the vascular tissue of raspberries. This disease typically causes wilting, stunting, and death of fruiting canes, while severe infections may cause death of the entire plant. Sapwood of infected canes is often stained red-brown. Verticillium also persists in the soil for many years.

Fireblight, although uncommon and of little concern on raspberry, can cause fruit loss and shoot death. Canes of infected plants first appear water soaked or dark green in colour. Canes may later turn purple to black and curl at the tip (Shepherd's Crook). Fireblight, a bacterial disease, may ooze from cracks on infected canes.

Raspberry leaf spot occurs on the upper surface of the leaf as white, gray, or tan spots. Sometimes the centre of the lesion falls out leaving a shot-hole appearance to the leaf. Heavily infected leaves may fall prematurely, increasing the plants susceptibility to winter injury. Leaf spot symptoms are often confused with similar symptoms of raspberry anthracnose. To differentiate these two diseases, examine the stems closely for lesions, which indicate the presence of raspberry anthracnose.

Powdery mildew usually appears on the bottom side of leaves as a white to gray powdery growth. This disease may occur yearly, but is not considered a serious problem. Rust fungi appear on the upper and lower leaf surface as orange-yellow spots.

Several rust fungi commonly infect the leaves and canes, while only a few fungi are able to infect fruit. Rust fungi are fairly common, but do not typically affect the overall health of the plant. However, one fungus, commonly known as orange rust, is considered a serious threat to black raspberries. Red raspberries are immune. Orange rust is a systematic disease that survives winter in newly formed roots, causing new infections the following year.

Fungal and Bacterial Disease Management

To properly manage fungal diseases, begin with the cultural practices previously listed. Fungicide control is available for anthracnose, cane blight, and spur blight. A dormant season application of lime sulfur can be applied in early spring as the buds begin to swell, but before leaves appear. Sulfur and copper-based fungicides are the only other chemicals available for homeowner use. Most of the remaining diseases do not require fungicide applications. To control severe verticillium infections, destroy infected plants and establish a new stand in a different location. To control fireblight, prune out and destroy infected canes. Sterilize pruning tools in between each cut with a 10% bleach solution or 70% alcohol. To control orange rust, remove and destroy infected raspberry plants. A laboratory examination is recommended before removing infected plants.

Viral Diseases

Viral diseases, although less prevalent than fungal diseases, are considered more damaging. Viral diseases often limit both the yield and longevity of a raspberry planting. The most common viral diseases found in Minnesota include raspberry mosaic, leaf curl, and tomato ringspot. Raspberry mosaic disease typically causes yellow-green leaf mottling, leaf blistering, leaf tip browning, and/or stunting. Raspberry mosaic disease can also infect plants without expressing any visual symptoms.

Raspberry mosaic disease is caused by a complex of viruses and transmitted by the large raspberry aphid. Severe infection causes a reduction in stand vigor, fruit quality, winter hardiness, and stand yield. Raspberry leaf curl virus typically causes curling and distortion of dark green leaves. Infected plants appear stunted and contain excessive branching. Infected fruit is usually small, crumbly, and seedy. This disease is spread slowly by the small raspberry aphid. Severe infection causes a reduction in yield, fruit quality, and winter hardiness.

Tomato ringspot virus is carried to healthy plants through the soil by the root feeding dagger nematode (Xiphinema sp.). Yellow rings, line patterns, or veinal chlorosis commonly appear on the leaves of red raspberry plants. However, new leaves are often void of symptoms. Severely infected plants may appear stunted, yield crumbly fruit, produce poor yields, and/or show a reduction in plant health. Tomato ringspot virus can also be dispersed through infected seed.

Model Farming

The viruses are parasitic in nature and cause the most infectious group of diseases, of all forms of livings (including human being, animals and plants). The viruses are very small to that of all others and can not be seen with naked eye, nor with the help of commonly used microscopes, hence are regarded as sub microscopic, nucleo protein particles, multiply inside living cells. Viruses cause diseases by upsetting the metabolism of the cells, but not by consuming cells or killing them with toxins. The total number of viruses known to date is well over a thousand, and new viruses are added to this almost every month. More than half of all known viruses attack and cause diseases of plants. One virus may infect one or dozens of same or different plant species, and one plant may be attacked by one or many different viruses. The importance of plant viruses in relation to crop production, can be realized from the fact that among the various factors responsible for low yields, viral diseases are prominent and cause losses in world's crop production amounting to many million rupees, which comes next only to losses caused by insect pests. Plant virus diseases may damage leaves, stems, roots, fruits, seed or flowers and may cause economic losses by reduction in yield and quality of plant products. The severity of individual virus diseases may vary with the locality, the crop variety, and from one season to the next. On nation wide bases, the record showed that amongst the major viral diseases of economic crops, recorded in Pakistan (Table), some virus diseases have destroyed entire planting of certain crops in some areas, for example, cotton leaf curl virus, banana bunchy top virus, viral diseases of chilies, tomatoes and pulse crops are considered very serious, during different cropping seasons, at different locations of Pakistan; and in certain cases, fields have been found to show as much as 100 percent damage. However, it is quite difficult to present accurate estimates of the losses due to viral diseases. It is becoming obvious that: now, it is upon the crop protectionists including plant pathologists and entomologists as well, to design and formulate ways or mean to combat all enemies of the crops, so that the growers (farmers) may try to minimize the losses caused by plant viruses to the crops.

Identification of Viral Diseases

Various external and or internal changes, reactions or alterations (signs) on or in the plants, due to any abnormality as a result of the pathogen (disease causing agent) is termed as symptoms. Actually, abnormal appearance on or in the plant is usually the first indication of a virus infection in nature. Severe disease symptoms may occur only when virus has infected the plant systematically. It must be remembered that a virus not only causes just one type of symptoms. Mostly viral infection results in more than one type of symptoms. There may be a series of symptoms as the disease persists within the plant. For example, stunted growth and dwarfing, may be associated with necrotic symptoms and in extreme cases, the necrosis may spread to the whole plant to cause plant death.

Most Common External Symptoms

The most obvious symptoms of virus infected plants are usually those appearing on the foliage, but some viruses may cause striking symptoms on the stem, fruit, and roots, with or without symptom development on the leaves. The most common types of plant symptoms produced by virus infections are mosaic, mottle, vein clearing, vein banding, yellows, ring spots, chlorosis, dwarfing and stunting, tumours or galls, bunchy top, witches broom, rosette, enation and necrosis.

Mosaic: Mosaics characterized by intermingled patches of normal and light green, yellow or white areas of the leaves or fruits, or are whitish areas intermingled with areas of the normal colour of flowers or fruits. The mosaic depends on the intensity or particular pattern of discoloration. The mosaic type symptoms may be described as mottling, streak, ring pattern, line pattern, vein clearing, vein banding, chlorotic spotting, etc. The viruses causing most mosaic diseases are mechanically transmitted and usually have aphid vectors in nature, are generally resistant to brief heat treatments, and do not stop flowering or effect the dormancy of buds.

Mottle: An irregular pattern of indistinct light and dark areas.

Vein Clearing: Veins become clear due to destruction of chlorophyll in the vein tissues.

Vein Banding: Bands of green tissue along the vein, while the tissues between vein become chlorotic.

Yellows: When chlorophyll disappears completely due to chlorosis, yellowing, bronzing or reddening, the foliage of the host becomes uniformly discoloured without any spotting patterns and become yellow, although some vein clearing may be present. Viruses causing the true yellows diseases show a tendency to produce virescent flowers, to break the dormancy of axillary buds and induce cessation of flowering, to be leafhopper transmitted, and to be relatively sensitive to heat treatment.

Ring Spots: Ring spots, characterized by the appearance of chlorotic or necrotic (usually circular) ring spots on the leaves and sometimes also on the fruit and stem. Most ring spot causing viruses are not transmitted by either aphids or leafhoppers, but some of them are transmitted by nematodes.

Chlorosis: Yellowing of green tissues due to chlorophyll destruction.

Dwarfing And Stunting: The plant size is reduced due to shorter internodes, smaller leaves, fruits and various other plant parts.

Tumour (Galls): Unusual swelling or development or transformation produced as a result of viral infection.

Bunchy Top: Leaves or branches become bunched at the top of plants.

Witches Broom: Appearance of broom like growth, due to pathogen.

Rosette: Short, bunchy habit of plant growth.

Enation: Malformation or tumour or leaf like out growths on the leaves and roots referred to as enation.

Necrosis: Death of cells or tissues.

Less Common Symptoms: A large number of other less common virus symptoms are also described.

These symptoms may be accompanied by other symptoms on other parts of the same plant and include: leaf roll, leaf and stem distortion (e.g., bean common mosaic virus BCMV), rubbery wood, pitting of stem, pitting of fruit (for example, pear stony pit) and flattening and distortion of stem.

Cotton

Cotton leaf curl is recorded as most destructive diseases, while, sometimes stenosis (stunt or small leaf) also become important.

Cotton Leaf Curl Virus (CLCV)

This disease is also called leaf crinkle. A virus causing leaf curl of cotton was first recorded in Nigeria (1912), Sudan (1924), Tanzania (1926), Philippine (1959). In Pakistan, this disease was first time recorded in 1967 at Multan (Punjab) on some cotton plants.

It was considered a miner disease until 1987, but in 1991 92, it becomes severe and since 1992 93 causing a huge production and monetary loss to the nation. In Sindh, this disease was first reported during 1996 at Ubauro, district Ghotki, and is reached up to New Saedabad, district Hyderabad, during 1999 2000. It is quite difficult to present accurate estimates of the losses due to cotton leaf curl disease, because the losses vary from year to year and from one area to the other. Sometimes the cotton fields have been found to show as much as 100 percent damage.

Pathogen: The disease causing virus belongs to Gemini group, sometimes refer as Gossypium virus 1.

Distribution: Africa, Nigeria, Sudan, Tanzania, Philippine and Pakistan.

Host Range: More than 30 different crop, weed and ornamental plants are reported as hosts.

Seriousness: Cotton, lady's finger, tomato, chili, cucurbit (especially water melon), beans, sunflower, sesame, soybean, cow peas, egg plant (brinjal), sun kukra, china rose, thorn apple (dhatura), mint (podina), holly hock (gul e khera), zinnia, AK (Calotropis), shesham (talhi) and citrus species.

Transmission: The disease transmitted by feeding of the white fly, Bemissia tabacci within 6.5 hours. A single female, carrying virus, can infect many plants. White fly is known to survive on as many as 53 host plant species, and is responsible for transmitting 23 crop diseases in region. At global level, white fly infests 600 different plant species. The virus is not transmitted by sap, seed or soil.

Perpetuation: The disease causing virus survives in several different plant hosts, from which it may spread.

Symptoms: Upward and downward curling of leaves accompanied by small and main vein thickening (SVT & MVT) on leaves, pronounced on underside. If a diseased leaf is viewed from beneath against the light, thickened vein found darker green and opaque than normal. In extreme but not in frequent cases, formation of cup shaped or leaf laminar (veins) outgrowth called "enation" appears on the back or underside of the leaf. The newly produced leaves are small, excessively crinkled and curled at the edge. Primary stem often tends to grow taller than normal. The internodes being elongated and irregularly curved but sometimes the whole plant is stunted. The flowers checked in growth and become abortive. Bolls remained small in size and failed to open. All parts of badly hit plants are very brittle and ready broken.

Control: Cultivation of resistant varieties is only safe measure. Crop rotation with non host crop. Proper use of irrigation and fertilizers. Potassium fertilizer improves the disease resistant power in plants. Vector, white fly must be controlled. All alternate hosts (including weeds) must be eradicated before, during and after cotton crop. Deep plowing with short duration in fallow lands help to control weed hosts. The disease (CLCV) is not seed transmitted but use of healthy seed, acid delinting and chemical seed treatment is recommended as preventive measure. Use of proper cotton production technology is economical and most effective for management of all diseases (including this).

Stenosis (Stunt or Small Leaf)

Distribution: Pakistan and India

Symptoms: Leaves develop in clusters, are malformed and of different shapes and sizes. Enations are produced on lower surface of veins. Flowers may remain small with balls never forming. Infected plants can easily pulled out of the ground, having a large number of adventitious roots.

Sugarcane

Mosaic, grassy shoot and chlorotic streaks are considered major viral diseases of sugarcane.

Mosaic

Pathogen: Sugarcane mosaic virus (SMV).

Distribution: Wherever sugarcane is grown.

Host Range: The disease causing virus has a wide range and infects a large number of grasses.

Transmission: Aphids, Mechanical, and is seedborne in corn.

Perpetuation: Grasses and infected sugarcane crop.

Symptoms: Newly leaves are unrolled from spindle. Irregular oval or oblong, pale green blotches of various sizes occur on leaves, with various widths. Stunted shoots, twisted and distorted leaves in some cultivars. Mottling of stem, causing death of tissue and cankered areas in other cultivars may also occur.

Control: Plant resistant varieties. Rogue out infected plants.

Grassy Shoot

Pathogen: Sugarcane grassy shoot virus (SGSV)

Distribution: Pakistan, India, Taiwan and Thailand.

Host Range: Sugarcane and sorghum.

Transmission: Infected sugarcane seed pieces, mechanically by cutters or cutting knives and aphids.

Perpetuation: Infected sugarcane crop.

Symptoms: Typically, mass of stunted, crowded shoots arise from diseased seed pieces or ratoon stools. Some shoots maybe devoid of chlorophyll and stool may die. Stalks may have long, delicate, scaly buds that develop chlorotic stools. White to yellow, well defined stripes occur on thin and chlorotic leaves. Numerous tillers may also occur at base of the stool.

Control: Plant disease free cuttings.

Chlorotic Streaks

This disease is most prevalent in low lying, poorly drained areas and potassium deficient soils.

Pathogen: Sugarcane chlorotic streak (SCSV).

Distribution: Pakistan, U.S.A., Australia and Java.

Host Range: Several grasses.

Transmission: Stalk cuttings, leaf hopper and running water.

Symptoms: One to many yellowish to whitish streaks with wavy, irregular margins occur on both sides of leaves, leaf midribs and sheaths. Streaks are short at first but later on may extend the entire length of leaf blade. Erect leaves and wilt in young plants. Young shoots may stunt or die.

Control: Plant resistant varieties. Hot water treatment at 52OC. Use disease free material. Rouging of infected crop plants and weeds. Improve soil drainage. Proper application of potassium fertilizers as per requirement.

Wheat and Barley

Yellow Dwarf is considered most important viral disease of wheat and barley, but in Pakistan, is not recorded as destructive.

Pathogen: Barley yellow dwarf virus (BYDV).

Distribution: Africa, Asia, Australia, Europe, New Zealand, North and South America.

Host Range: Wide variety of gramineous plants.

Seriousness: Barley, oats, wheat, rye, lawn weeds and gasses.

Transmission: Dodder and 11 species of aphids and 6 species of hoppers.

Perpetuation: Through cereal and grass hosts as well as aphids.

Symptoms: Yellowing and shortening of leaves, stunting of plants, reduced tillering but severely infected barley plants may show excessive tillering. Sometimes the head fails to emerge. Inflorescences are smaller and number and weight of kernels is reduced. Many of the flowers are also sterile. The root system is drastically reduced in weight but shows no symptoms.

Control: Cultivate resistant varieties. In some cases, early or late sowing and high seed rate may also be helpful. Control of vectors.

Maize and Sorghum

Streak Disease has been reported as viral disease but has not importance with reference to damaging the crop.

Distribution: Asia and Africa.

Host Range: Maize and sorghum.

Transmission: Through leafhoppers (not seed or sap).

Symptoms: Initially, circular, colourless spots occur on lowest exposed portions of young leaves. Spots are scattered at first but later on become closer. Narrow, broken, chlorotic stripes occur along veins. The stripes may coalesce to form wider stripes.

Bean Crops

Mosaic is considered as major threat to almost all BEAN CROPS.

Pathogen: Bean yellow mosaic (BYM) and Common bean mosaic (CBM).

Distribution: Generally wherever field beansare grown.

Transmission: Yellow mosaic is transmitted through aphids only but common mosaic is seedborne, may also transmit through aphids, pollens and by mechanical mean.

Perpetuation: Wild sweet clover helps the virus to survive.

Symptoms: Differences between symptoms may vary greatly between plants. However, in both diseases, the general symptoms include dwarfing, excessive branching or bunches, leaf cupping and typical symptoms of mosaic.

Control: Elimination of wild sweet clover is recommended for yellow mosaic. Disease free seed is suggested for common mosaic. Resistant varieties or rouging of infected plants are best way against both diseases.

Groundnut

Rosette is alone important viral disease of groundnut.

Pathogen: Groundnut rosette virus (GRV).

Distribution: Africa, Java and Pakistan.

Symptoms: An overall stunting of plants with typical rosette or clumped appearance is common symptom. Affected plants have flattened growth at the top portions associated with leafy growth and malformed buds. Young leaflets become faint in colour followed by chlorosis. Chlorotic and mottled leaf, and blossom and pod formation is reduced. Early infection causes small, sessile flowers that do not open.

Control: Sow resistant varieties. Diseased plants should be uprooted and destroyed.

Sesame

Leaf Curl is recorded as problematic disease.

Transmission: White fly, Bemissia spp. is the main vector in nature, and graft transmission is reported successful for producing the disease.

Symptoms: Diseased leaves are markedly reduced in size, become slightly thick and brittle and dark green in colour. Affected plants are stunted and bear scanty capsules having poor seed setting. Early infection may result in severe reduction in yield.

Control: Diseased plants and weeds must be collected and destroyed. The vector white fly may be controlled.

Tomato and Tobacco

Mosaic has equal importance as very important viral disease of tomato and tobacco crops.

Pathogen: Tobacco mosaic virus (TMV)

Distribution: World wide.

Host Range: The virus may affect more than 150 genera of primarily herbaceous, dicotyledonous plants including many vegetates, flowers and weeds.

Seriousness: Tobacco and tomato show symptoms, but symptomless on grape and apple.

Transmission: Contaminated hands, tools or equipments. Nursery plants, Sap, Grafting and Dodder, Seed transmission in apple, pear and grape. Occasionally by contaminated jaws and feet of insects.

Perpetuation: Through plant debris, seed, cigarettes and snuff. The viruses retain infectious in dried mosaic infected leaves when heated at 120OC for 30 minutes, at dilution of 1:1000000. Remain infectious for more than 50 years in TMV infected leaves kept dry in laboratory.

Symptoms: Consist various degrees of: Chlorosis, curling, mottling, dwarfing, distortion and blistering of leaves, Dwarfing of entire plant, Dwarfing, distortion and discoloration of flowers. In some plants, development of necrotic areas on the leaf. TOMATO: Mottling of older leaves, mottling with or without malformation of leaflets. Long, pointed and sometimes shoestring like leaflets. Reduced fruit set.

Control: Sanitation, Resistant varieties, 2 years crop rotation, removal of infected plants, weeds and alternate hosts. Avoid chewing and smoking tobacco during handling crop or washing hands with soap. Spray milk on plants and dipping the hands in milk during handling the crop.

Tomato

Leaf Curl is considered as secondarily important viral disease of the tomato crop that may also cause considerable loss in quality and quantity of produce.

Pathogen: Tobacco virus 16 or Nicotiana virus 10

Transmission: White fly, Bemissia tabacci.

Symptoms: Dwarfing, twisting and curling of leaves, mottle vein clearing, excessive branching, stunting of plants and partial or complete sterility.

Control: Planting resistant varieties.

Chili

Leaf Curl is considered most important viral disease of chili after wilt caused by fungi.

Pathogen: Nicotiana virus 10

Transmission: Through graft or white fly, Bemissia tabacci.

Symptoms: Curling of leaves, accompanied by thickening and swelling of veins. Clusters of leaves with reduced size. The whole plant assumes bushy appearance with stunted growth. Fewer flowers and fruit, but if are formed, are much reduced in size and are curled.

Control: Planting resistant varieties.

Lady's Finger (Bhindi or Okra)

Yellow Vein Mosaic is considered very important disease.

Pathogen: Hibiscus or Bhindi yellow vein mosaic virus.

Transmission: White fly, Bemissia tabacci.

Symptoms: Vein clearing, vein chlorosis, yellow veins enclosing green patches of the leaf. Veins are thickened on lower surface of the leaf. Fruits are develop malformed and reduced in size, mostly are yellow, small, tough and fibrous.

Control: Planting resistant varieties. Eradication or rouging of infected plants and weeds.

Cucumber (Cucurbit)

Mosaic is recorded as major problem in almost all members of cucurbit.

Pathogen: Cucumber mosaic virus (CMV).

Distribution: World wide.

Host Range: Wide range of hosts than any other virus.

Seriousness: Cucurbit, peppers, spinach, tomatoes, beets, beans, banana, crucifers, lilies, zinnia and many weeds.

Transmission: Sap and several insects. Mostly 6 spp. of aphids, including Myzus persicae and Aphis gossypii. Not through seed in cucumber but is in some hosts.

Perpetuation: Through weeds, flowers and crop plants.

Symptoms: Cotyledons turn yellow and wilt. Young leaves become mottled, distorted, wrinkled and their edges begin to curl downward forming rosette like clump. The plants appear dwarfed, leaves become half to their normal size. Few runners, flowers and fruits. Older leaves develop chlorotic and necrotic areas that cover entire leaf and killed leaves hang down or fall off. Fruit shows pale green or white areas intermingled with dark green, raised areas. Often form rough, wartlike projections, cause distortion, are somewhat misshapen but have smooth gray white colour with some irregular green areas. Often have bitter taste and upon picking become soft and soggy.

Control: Resistant varieties. Use of disease free seedlings. Elimination of alternate hosts (including weeds). Control of insect vectors.

Citrus

Tristeza Disease of Citrus is an important viral disease of about all citrus plants or trees.

Pathogen: Citrus Tristeza virus (CTV).

Distribution: South America, South Africa, Asia and Australia.

Host Range: All kinds of citrus plants or trees.

Seriousness: Orange, lemon, grapefruit, limes etc.

Transmission: Rootstocks, buds, grafts, dodder and aphid.

Perpetuation: Infected hosts, sometimes 100% trees of an orchard carry the virus.

Symptoms: Symptoms vary on the citrus species. More or less sudden wilting and drying of the leaves followed by death of tree or dieback of twigs and partial recovery may occur. Leaves become small, more or less yellow, stand upright. Death and decay of feeder rootlets and extends to primary root.

Control: Use of disease free rootstocks, buds and grafts. Removal of infected trees.

Banana

Banana Bunchy Top is considering a very severe disease, causing a huge production and monetary loss to banana growers in all over Sindh. The disease was first recorded from Fiji in 1879, in India during 1940. Now this disease is prevalent also in Pakistan. The symptoms of banana bunchy top disease were first time recorded in December 1988 at coastal areas of district Thatta, Sindh, but the disease was confirmed during July 1991.

Pathogen: The bunchy top of banana is caused by banana bunchy top virus (BBTV) Magee.

Distribution: Banana bunchy top disease is widely distributed among banana growing countries viz. Australia, Fiji, Egypt, Sirilanka, Bunion, Island, Pakistan and India.

Host Range: No alternate host has been reported.

Transmission: The disease causing virus is being transmitted by an insect victor, black aphid, Pantalonia nigronervosa Coq. The aphids generally live round the base of pseudostem. The virus is not transmissible by mechanical mean, but is systematic, and found on all parts of the infected plants. Suckers are considered a major source for spread of the disease.

Perpetuation: The disease perpetuates through infected plants and virus migrates through infected plant parts, suckers and vector.

Symptoms: The symptoms may appear at any stage of banana plant growth. The leaves become bunched together at the apex of plant and a rosette is formed. All the symptoms produced could be categorized in following distinct groups.

Stunted and erect leaves: Emerging leaves are progressively shorter, narrower and more erect than normal.

Stunted plants: Infected plants do not usually grow taller than two to three feet, hence are markedly stunted.

Dot dash lines on leaves: The irregular, nodular, dark green dot dash lines or streaks appear running down the leaf stalk between the leaf blade and pseudostem.

Colour, thickness and size of leaves: Diseased leaves become pale yellow, brittle and much reduced in size.

Small, deformed bunch and bell: Only on plants, that are infected late, bunches and flower bells are stunted and deformed. Often the bunch stem will curve outward.

Dot dash lines on bell: Dark green dot dash lines occur on the bottom of the flower bell. Sometimes the flower bracts bend upward.

Control: No resistant varieties have been evolved, hence following preventive measures must be adopted. Regular visit of banana may help to sort out diseased suckers and plants for eradication. Quarantine regulation must be implemented, so that the diseased suckers and other plant parts should not be transported to disease free areas. Certified, disease free suckers should be planted. Extra new suckers must be removed. Row to row and plant to plant space may be maintained. Insect vectors that is aphid be controlled. Weeding may be confirmed during interculturing with short interval. Proper use of irrigation and fertilizer could also be helpful. Modern production technology is safe way to save the crop.

Papaya

Leaf Curl is important viral disease of papaya.

Distribution: Where papaya is grown.

Transmission: White fly, Bemissia tabacci.

Symptoms: Almost all the leaves of the plant are reduced in size and show malformation and sever curling, crinkling and distortion. The margins of the leaves are curved or rolled downward and or inward. Vein clearing and thickening also take place. The leaves become brittle and growth of the plant is arrested. Sometimes, plants become partially or completely sterile depending on growth stage and severity of the disease. If the fruits are formed, these are disfigured and mummified.

Control: Planting resistant varieties. Eradication of diseased plants.

A Plant Disease Primer

Healthy gardens teem with microorganisms: bacteria, fungi, viruses, virus-like organisms, and nematodes. The majority of these are either benign, coexisting in the soil and on plant surfaces with each other and with the plants around them, or they are beneficial, and perform such tasks as breaking down the compost pile into humus, transforming the nitrogen in the air into forms that plants can use, or preying on pest species. Only a small percentage of microorganisms are pathogenic—that is, capable of causing plant diseases. Fungi are the cause of the largest number of plant diseases, while viruses often cause those with the most severe symptoms.

Many pathogens cause only cosmetic or superficial damage. But a few of the most virulent kill their hosts outright and can pose a problem in the garden for years to come.

Plants can also develop disorders. These are not brought on by pathogens, but rather by environmental conditions and stresses. Common examples include blossom end rot of tomatoes and peppers, and tip burn of lettuce, both of which are caused by a calcium deficiency in combination with fluctuating moisture conditions and fluctuating light levels, respectively. In most cases, the symptoms of disorders are confusingly similar to those resulting from pathogenic diseases.

Plant Defenses

Unlike animals, plants don't have an immune system or disease-fighting white blood cells. Instead, they have developed other mechanisms to escape or survive the diseases that attack them. Some plants are simply genetically incompatible with, or immune to, certain diseases. A corn plant can't host tobacco mosaic virus (TMV), a common tomato disease, for example, just as tomatoes can't host the corn disease smut. It is also possible that a particular cultivar, variety, or even an individual plant is immune to a disease that attacks other members of its species, generally because of subtle differences in the genetic makeup of the various plants. Resistance to diseases is another factor that protects plants. Although a resistant plant can be susceptible to a particular disease, it is less likely to become infected by it than a non-resistant plant. In some cases, plants are resistant because of a physical characteristic: for example, their cuticles may be too thick for a particular fungus to penetrate, or their stomata, small pores on the leaves, may be closed at the time of day when environmental conditions stimulate disease spores to germinate. Other types of resistance are biochemical: the plant produces enzymes and proteins that inactivate the pathogen.

A plant is described as tolerant to a particular disease if it can be infected by the disease but is unlikely to be killed or severely disabled by it. Tolerance often results from hypersensitivity: a plant may be so sensitive to a disease organism that the cells infected by the pathogen die very rapidly and the disease organism is isolated and cannot move to other areas of the plant tissue.

Plants can also escape infection. Some diseases, such as early blight of tomatoes, primarily attack plants that are young and growing vigorously, while others, such as late blight of tomatoes, attack those that are old and forming more fruit and seeds than new leaves. So plants that are older when early blight strikes have a better chance of surviving, as do those that are young when late blight attacks. Similarly, some pathogens, such as many that cause powdery mildew, slowly build to damaging populations as the season progresses, while others, such as Pythium, which can cause root rot, are at their most numerous in the early spring. That is to say, it may pay to put off planting crops susceptible to root rot, such as potatoes, until later in spring when the soil has begun to dry, and you may want to take measures to ensure that you can harvest your cucumber crop early, before powdery mildew becomes a problem. When plants do not become infected with a disease simply because of timing, they have "escaped infection."

Disease-causing Organisms

Bacteria

Bacteria cause three types of plant disease symptoms. Some live within a plant's vascular system, the channels through which water and nutrients move from roots to leaves and vice versa. When these organisms become numerous, they are likely to plug up the system, preventing the flow of water and nutrients. Wilts, often beginning on only one side of the plant or even on one branch, are the result.

Other bacteria cause abnormal growth in the cells they have invaded and/or in adjacent cells. These growths, known as galls, may pepper the surface of a leaf or protrude from a stem. They are usually rounded and hard to the touch. When bacteria kill the cells they infect outright, the results can be rots, internal blights, or spots. Sometimes the spots are small and barely noticeable, but in other cases, cankers, open wounds, or lesions form on bark or soft tissue. When the infection is serious, the bacteria spread and kill the entire leaf, stem, or plant.

Most bacteria are one-celled organisms, although a few are multi-cellular. They reproduce by cell division. Because this reproduction is not sexual, bacteria cannot hybridize. However, that characteristic does not inhibit their ability to evolve; they do exchange genetic information and DNA through plasmid exchange (the exchange of small, DNA-containing elements that exist outside the chromosome), and they mutate relatively easily. Changes in their environment, including ultraviolet light levels and ambient levels of atmospheric gases, as well as subtle changes within their host, can trigger a mutation. This adaptability is responsible for the steadily increasing number of bacteria that cause disease in plants and animals—and the increasing number of bacteria that are immune to antibiotics.

When conditions such as temperature or moisture levels do not suit them, bacteria become dormant. Although many overwinter in the seeds of infected plants or inside insects (the pathogen that causes bacterial blight of cucurbits overwinters in the salivary glands of hibernating cucumber beetles, for example), most overwinter on cankers on dormant plants or in the debris of the plant they infected during the growing season. Though a few species can survive for longer than a year in a dormant state, most die after this time. Consequently, rotating crops so they don't grow in the same area again for two to four years is usually an effective control.

Fungi

Fungi are far more complicated in both structure and reproductive style than bacteria. The fungal body, called a thallus, usually consists of microscopic, threadlike hyphae that spread throughout the source of food of the fungus. Collectively, these hyphae are referred to as the mycelium. The reproductive strategies of different species of fungi vary widely. Some form sclerotia, thickened areas of a hypha, which can break away or overwinter in place and grow into a new organism.

Other fungi form spores, tiny bodies that can germinate and grow as seeds do. Some spores are formed asexually; these are usually known as summer spores because they can reproduce as soon as they find appropriate conditions. They tend to die within a few months if they do not find a hospitable environment.

Other spores are formed sexually. Many of these are called resting spores because they can remain dormant but viable for months or, if necessary, for many years. Almost all fungal species produce at least two kinds of spores to guarantee their survival, but the fungal rust diseases that require two alternate hosts produce as many as five spore forms over the two years it takes to complete their entire life cycle.

Fungi mutate in the same way that bacteria do; and those that reproduce sexually have the additional advantage of being able to crossbreed. Like bacteria, fungi are capable of evolving rapidly. Pathogenic, or disease-causing, fungi produce symptoms that are similar to those caused by bacteria: wilts, galls, rots, spots, cankers, and blights, as well as smuts (large growths filled with spores), scabs (roughened skin tissue), and mildews. These last three symptoms are characterized by the visible presence of the hyphae.

Fungi overwinter in plant debris, seeds, or the soil, as spores or sclerotia. Depending on the species, resting spores can remain viable for two to 20 years. It pays to do some research before simply assuming that a crop rotation will protect your plants from infection by a fungus.

Viruses

Viruses that attack plants are just as insidious as those that attack animals. Their physical structure is simple: just a bit of DNA or RNA and a protein covering. Rather than reproducing themselves, viruses stimulate the cell within which they are living to create more viruses. They too evolve quickly in response to changes in their environment.

Viral diseases cause malformations, stunting, and disruption or death of cells containing chlorophyll, which often results in the yellowing of plants (chlorosis). The malformations range from simple leaf curling or puckering to abnormal production of stems or branches, as in the disease called witches' broom. Stunting can be generalized, involving the whole plant, or specific to a particular area. Mosaic diseases are characteristic of viruses that interfere with chlorophyll production. Mosaic-infected leaves are usually mottled with irregularly shaped yellow patches.

Viruses can remain in the host without being active. Many have a fairly narrow band of temperature preference; they are active between 55°F. and 85°F. but not at temperatures above or below this. This characteristic often gives gardeners the false impression that their plants suddenly "got well" from a viral disease.

Plant debris, soils, and gardeners' tools, hands, and shoes can all transport dormant viruses from one host to the next. But viruses remain viable for so long—up to 50 years in some cases—that it is impossible to know for certain how a virus got into a planting. They can also survive the hottest temperatures a compost pile can achieve.

Nematodes

Nematodes are microscopic animals that look like slender, nonsegmented worms. There are thousands of nematode species, but only a few hundred of them cause plant diseases. These are directly responsible for a number of symptoms, including rotting tissues and abnormal growth, but they also cause diseases indirectly. They can carry viruses, injecting them into the plant as they feed. Moreover, when nematodes pierce the plant tissue to move into cells, secondary pathogenic organisms very often take advantage of the openings to invade the plant.

Nematodes reproduce both sexually and parthenogenetically—that is to say, without the benefit of males. They produce eggs, which are released into the soil around a plant or inside the plant itself.

Eggs develop into juveniles, the stage at which most nematodes infect plants; juveniles can also remain dormant but viable for months without a host. Nematodes overwinter in soil or plant tissue as eggs, juveniles, or adults. Egg-to-egg life cycles usually range from three weeks to three months.

Spreading Disease

Pathogens move from host to host in a number of ways. When not seed-borne, bacteria are usually carried by animals, insects, moving water, air currents, and infected soil, or spread around the garden on tools and shoes. Bacteria move into plants through natural openings, such as stomata, or through small wounds and bruises. Once inside a susceptible, living host, their environmental needs are met. Bacteria reproduce within a host for as long as it continues to feed them.

If the bacteria are feeding on the surface of a plant or kill enough cells so that the plant tissue breaks open, they are easily moved from one host to another by wind, falling rain, insects, or even gardeners. Those that feed internally usually overwinter on the plant debris or in the soil, ready to be carried to a new host the following spring.

Fungal spores travel the same routes from host to host as bacteria, but differ in that a spore must germinate and grow before it can infect a plant. Germination is primarily dependent upon temperature and humidity levels. Most fungal species require moderate (50°F. to 90°F.) temperatures and very high humidity or a film of water on the leaves.

After germination, most hyphae grow and find their way into a plant through natural openings, wounds, and bruises, but some species have an even more direct route. They form a penetration peg, a toughened tip on the end of the first hypha, that bores through the plant's cuticle layer and into the tissue.

Other species release a toxin that kills plant cells; once surface cells are dead, they move into the plant's interior, always killing cells in advance of their spread.While some fungi can feed only on dead material, and some can feed only on living plants, many of the plant pathogens can feed on either dead or living material. This enables them to remain active after their host is dead, as long as temperatures are tolerable for them; once the weather warms in spring, they can start feeding on the host again and produce a new crop of summer spores that are ready to infect new plants.

Viruses are usually transmitted by insect or nematode vectors—organisms that carry and transmit a disease—and gardeners' hands and tools. Vectors generally transmit the disease by inserting it into the plant as they are feeding. Gardeners are more likely to infect a plant by touching it with dirty hands. Viruses are so small that they can be moved into wounds and bruises that are invisible to the naked eye. As long as the host is alive and susceptible to them, their environmental needs are met. Nematodes that infect plants do so in their larval stage. If they are not carried to new hosts by moving water or on shoes or tools, they swim through the soil solution. However, due to their size, even the speediest nematode rarely travels more than a foot away from its birthplace. Again, the only requirement for infection is that a susceptible host is alive.

Disease Causing Agents in Plants-Symptoms and Control

The formulation of economic and effective disease control strategies requires a fundamental knowledge of the causes of disease, their life cycles, symptoms on host plants, and the principles of control measures.

Equally important is the ability to survey and quantify disease within a plant population, and establish economic disease threshold levels. This HortFACT is devoted to providing a basic knowledge on the causes of disease by biotic (living) and abiotic (non living) factors, symptoms, principles of disease control, and conducting disease surveys to quantify disease within a crop.

Causes of Disease

Diseases Caused by Biotic Factors

Disease development depends on the interactions between the pathogen, host plant and the environment. Disease can be caused by various agents (living or non living) either acting singly or in combination one with the another.

These agents fall into several groups. Fungi cause the greatest number of diseases, but disease can also be caused by bacteria or viruses.

Disease causing organisms are called pathogens. Diseases mainly occur when pathogens alter one or more of a series of physiological processes within the host plant, eventually manifesting as morphological changes commonly called symptoms. Sometimes the presence of the pathogen is more obvious than any morphological change in the plant. Visible growths of the pathogen are often called the signs of the disease (as seen in rusts and powdery mildews).

Symptoms

Different pathogens cause specific disease symptoms within a host. When only one particular pathogen is involved there is often a number of symptoms which together form a characteristic symptom-picture or syndrome to which a specific disease name is given.

For example, in downy mildew, initial symptoms on leaves is the 'oil spot', which then develops fluffy sporulation on the underside of the leaf. Later, infected leaves produce necrotic lesions.

Thus, the recognition of disease by symptoms comes largely with experience. However, symptoms generally can be grouped depending on the process by which they develop. These are as follows:

1) *Death or Necrosis*-Various terms describe the extent and shape of necrotic lesions, particularly on leaves: stripe for narrow, elongated lesions; scorch, scald, and blotch for indefinite areas which often become blanched and then brittle; and leaf spot for well defined lesions of limited extent. Sometimes the dead tissue of a leaf spot falls away and the symptom is called shot-hole.

2) *Abnormal increase in tissues*-Hypertrophy-an increase in the size of cells in the plant tissues. Hyperplasia-an increase in the number of plant cells.

   The common symptoms of these types are leaf curl, witches' brooms, galls, warts, cankers and scab.

3) *Abnormal decrease in size*-Hypoplasia-failure to attain normal size or development. In some diseases it is common to see an overall dwarfing or stunting of the plant or affected plant parts.

4) *Colour change*-Yellowing, or chlorosis, is a common symptom of disease often associated with tissues

surrounding a necrotic area. The chlorophyll may be degraded or fail to develop. Other colour changes may also be seen, such as development of red pigments (called anthocyanins), and silvering of leaves through the formation of air spaces as affected tissues break down. Viruses commonly cause these symptoms. Colour changes can also be caused by non living (abiotic) factors such as nutritional deficiencies and should not be confused with symptoms caused by pathogens.

5) *Wilting*-An interference with the normal movement of water within the plant can cause wilting symptoms.
6) *Transformation of plant organs or their unusual development*-In some plant species, infected flowers can bear rudimentary stamens. In the case of cereal smuts, the grain will be filled with fungal spores rather than starch. In brown rot diseases of stone fruit, infected fruit can be transformed into mummified fruit.
7) *Tissue disintegration*-This symptom is mainly associated with fruit infections and is termed rot (for example, Botrytis bunch rot). It may be accompanied by a release of cell fluids (leak). In some cases the cells may crumble to a powdery mass (dry rot).
8) *Gummation*-Some pathogens can cause excessive gum formation, particularly associated with tree infections. Bacteria commonly cause this symptom.

Diseases Caused by Abiotic Factors

Many plant disorders are not caused by living organisms (pathogens). These are commonly referred to as abiotic, non-biotic, environmental, or physiological diseases. Abiotic factors can cause short duration impacts resulting in injury (frost damage), or long term impacts resulting in plant debility, and can be considered as disease (e.g. imbalance of nutrients such as manganese toxicity in soils). It is important to be able to discern diseases caused by biotic factors from abiotic factors. The response to abiotic factors can vary depending on the plant species and their adaptability. The following are some of the factors which cause common abiotic disorders:

1) *Soil moisture*-Low soil moisture can cause the accumulation of toxic ions, such as manganese and boron, causing tissue damage. Other consequences are reduced photosynthesis and wilting when atmospheric demand is high. High soil moisture can have several adverse effects. Low oxygen in the soil can increase anaerobiosis and the build up of toxic products. These can initially cause root damage and eventually lead to shoot damage.
2) *Soil temperature*-Low soil temperatures usually stunt plants, while extremely low temperatures, can cause root death. Moderately low temperatures can prevent root development. High soil temperatures can also cause root death.
3) *Nutrient and mineral deficiencies*-Plant abnormalities and growth reduction can occur when there is a deficiency or imbalance in elements essential for plant growth. The deficiency symptoms can vary from species to species for a given element or nutrient. The following are some common deficiency symptoms:

   Nitrogen-loss of green colour, stunting, spindly growth and low yield.

   Phosphorous-reduced growth, although foliage remains green.

   Potassium-a variety of symptoms are caused including chlorosis, necrosis, and streaking, all showing up in older foliage.

   Calcium-deficiency first appears in the growing points of young leaves, causing distortion and death of young leaves. Growing tips fail to develop normally.

   Common examples are tip-burn of cabbage and blossom end rot of tomato.

   Magnesium-chlorosis firstly in the older leaves. Necrotic flecking or streaking are also characteristic symptoms.

Boron-poor growth of shoots and roots, and internal tissue breakdown.

Copper-deficiency symptoms can vary depending on plant species. On fruit trees a die back is seen.

Iron-chlorosis ranges from slight mottling to complete absence of chlorophyll.

Sulphur-similar to those of nitrogen deficiency.

Zinc-variable depending on plant species; little leaf in stone fruit, mottling in citrus.

4) *Chilling injury*-Some of the common symptoms of chilling injury are surface pitting, increased susceptibility to attack by weak and aggressive pathogens, and surface and internal necrosis.
5) *Freezing injury*-Usually freezing causes tissue damage to the internal cell contents. This can cause direct cell necrosis and even long cracks in the trunk of vines and trees.
6) *High temperature*-In general, high temperatures (>45°C) increase respiration, wilting, denaturation and coagulation of proteins, and cause injury and death of tissues in higher plants. Other possible symptoms are wilting, scorching of leaves and loss of leaves.
7) *Unfavourable light intensity*-Insufficient light may cause etiolation, stunting, reduced flowering, premature leaf fall and other abnormalities. High light intensities can scorch or roll leaves, dry up flowers, and accentuate spray damage.
8) *Wind damage*-Wind directly affects evapotranspiration, and its action can lead to root injury and in turn to root diseases. Wind can also carry spores of pathogens. Conversely, stagnant air can cause major air pollution problems. The symptoms of air pollution are dependant on the type of pollutant, and may result in plant toxicity leading ultimately to death.
9) *Improper use of chemicals*-Incorrect use of chemicals can cause serious damage to plants. These chemicals

include fertilisers, fungicides, insecticides, herbicides, stimulants, diluents, stickers and spreaders, and special action chemicals.

10) *Others*-Hail, ice and lightning can not only damage plants, but also produce wounds through which many pathogens may cause infection.

General Approaches to Disease Control

Disease control refers to the prevention of disease, or reducing/preventing the increase in the incidence and/or severity of disease. It is usually directed at plant populations rather than at individual plants. Satisfactory control of most diseases requires the adoption of multiple control measures, involving an integrated approach to manipulate environmental, biological, and chemical factors.

A well planned control programme should be based on knowledge of the characteristics of the pathogen and the host plant, the cultural and climatic conditions under which the crop is grown, and a knowledge of available disease control procedures including cultural, genetic, chemical, and biological approaches.

Plant disease control involves the application of one or more of the following principles:

1) *Avoidance*-by selecting areas where inoculum is ineffective, rare, or absent.

Avoidance should consider the following:

- choice of geographical area
- choice of planting site in a local area
- choice of planting time
- use of disease free stock
- modification of cultural practices.

2) *Exclusion of inoculum*-preventing the introduction of inoculum or the establishment of a pathogen within an uninvaded area. Pathogens may be excluded by:

- treating planting material
- using certified planting material

- restriction by quarantine
- eliminating vectors.

3) *Eradication*-reducing, inactivating, eliminating, or destroying inoculum at the source, either from a region or from an individual plant in which it is already established. Complete eradication is often difficult to achieve but some of the methods are:

- chemical control
- biological control
- crop rotation
- removal and destruction of diseased parts or whole plants
- elimination of alternate hosts and weed hosts
- sanitation.

4) *Protection*-preventing infection by applying a suitable toxic, or other effective barrier, between the susceptible host and the pathogen. Some of the common methods are:

- chemical
- biological
- controlling insect vectors of pathogen
- modification of the environment
- modification of nutrition.

5) *Disease resistance*-Breeding for resistance or tolerance to a disease can alter the effectiveness of inoculum through genetically altering either the physiological or structural characteristics of a susceptible host strain.

6) *Therapy*-reducing the severity of disease in an already infected plant. This involves one or more of the following methods:

- chemotherapy
- heat treatment
- surgery.

Control measures are predominantly preventative and should consider the interactions of host, pathogen and the environment. A plant species can grow under different environmental and climatic conditions. Therefore, different control measures must be selected for different environmental areas because of differences in cultural practices, temperature, humidity, inoculum potentials and virus-vector populations.

Disease Assessment-Measuring Disease

Disease assessment is the most important aspect in any crop loss programme because it is the process that generates data to quantify disease progress. It is critical to clearly define and standardise the disease assessment methods to avoid subjectivity and variability between assessors. Disease can be measured either directly, by measuring the disease on the plant, or indirectly, for example monitoring the production and presence of spores, or through remote sensing. The most common method of disease assessment is by measuring disease intensity. This refers to the incidence or severity of disease within a crop.

Disease incidence can be assessed with uniformity, while disease severity assessment has subjectivity. There are two methods for assessing disease severity:

1. A descriptive key which describes plants with different levels of disease and assigns category, number, index, grade or percentage to each description.
2. An assessment key which uses standard area diagrams which typify the development of disease on a whole plant or part of a plant.

It is advised that in both methods, a percentage scale be used for standardisation. Whenever disease assessments are recorded, it is also important to note the growth stage of the crop so that disease progress can be related to host development. It is not always necessary to adopt control measures the instant diseases appear in the crop.

Disease threshold levels need to be established for various diseases at different growth stages. Below these threshold levels, disease will not adversely affect plant growth or final yield. If disease increases to levels above set threshold levels, then suitable control measures need to be adopted. However, disease threshold levels have not been established for many diseases, and there is a need for research in this area.

Summary

In developing suitable disease control measures, the following steps need to be adopted:

1. Recognise the symptoms and/or signs of disease.
2. Identify the disease casual agent.
3. Quantify the disease using suitable disease assessment procedures (severity or incidence).
4. Determine threshold levels. If the disease is below threshold levels, then plot its progress over time. If the disease is above threshold levels, follow Step 5, and continue to monitor the progress of the disease over time.
5. Determine and implement relevant and suitable control measures.

Infectious Disease

An infectious disease is a clinically evident disease resulting from the presence of pathogenic microbial agents, including pathogenic viruses, pathogenic bacteria, fungi, protozoa, multicellular parasites, and aberrant proteins known as prions.

These pathogens are able to cause disease in animals and/ or plants. Infectious pathologies are usually qualified as contagious diseases (also called communicable diseases) due to their potential of transmission from one person or species to another. Transmission of an infectious disease may occur through one or more of diverse pathways including physical contact with infected individuals.

These infecting agents may also be transmitted through liquids, food, body fluids, contaminated objects, airborne inhalation, or through vector-borne spread.

The term *infectivity* describes the ability of an organism to enter, survive and multiply in the host, while the *infectiousness* of a disease indicates the comparative ease with which the disease is transmitted to other hosts.

An infection however, is not synonymous with an infectious disease, as an infection may not cause important clinical symptoms or impair host function.

Classification

Among the almost infinite varieties of microorganisms, relatively few cause disease in otherwise healthy individuals. Infectious disease results from the interplay between those few pathogens and the defenses of the hosts they infect. The appearance and severity of disease resulting from any pathogen depends upon the ability of that pathogen to damage the host as well as the ability of the host to resist the pathogen.

Infectious microorganisms, or microbes, are therefore classified as either *primary pathogens* or as *opportunistic pathogens* according to the status of host defenses.

Primary pathogens cause disease as a result of their presence or activity within the normal, healthy host, and their intrinsic virulence (the severity of the disease they cause) is, in part, a necessary consequence of their need to reproduce and spread. Many of the most common primary pathogens of humans only infect humans, however many serious diseases are caused by organisms acquired from the environment or which infect non-human hosts.

Organisms which cause an infectious disease in a host with depressed resistance are classified as *opportunistic pathogens*. Opportunistic disease may be caused by microbes that are ordinarily in contact with the host, such as pathogenic bacteria or fungi in the gastrointestinal or the upper respiratory tract, and they may also result from (otherwise innocuous) microbes acquired from other hosts (as in Clostridium difficile colitis) or from the environment as a result of traumatic introduction (as in surgical wound infections or compound fractures).

An opportunistic disease requires impairment of host defenses, which may occur as a result of genetic defects (such as Chronic granulomatous disease), exposure to antimicrobial drugs or immunosuppressive chemicals (as might occur following poisoning or cancer chemotherapy), exposure to ionizing radiation, or as a result of an infectious disease with immunosuppressive activity (such as with measles, malaria or HIV disease). Primary pathogens may also cause more severe disease in a host with depressed resistance than would normally occur in an immunosufficient host.

One way of proving that a given disease is "infectious", is to satisfy Koch's postulates (first proposed by Robert Koch), which demands that the infectious agent be identified only in patients and not in healthy controls, and that patients who contract the agent also develop the disease. These postulates were first used in the discovery that Mycobacteria species cause tuberculosis. Koch's postulates cannot be met ethically for many human diseases because they require experimental infection of a healthy individual with a pathogen produced as a pure culture. Often, even diseases that are quite clearly infectious do not meet the infectious criteria. For example, *Treponema pallidum*, the causative spirochete of syphilis, cannot be cultured *in vitro*-however the organism can be cultured in rabbit testes. It is less clear that a pure culture comes from an animal source serving as host than it is when derived from microbes derived from plate culture. Epidemiology is another important tool used to study disease in a population. For infectious diseases it helps to determine if a disease outbreak is sporadic (occasional occurrence), endemic (regular cases often occurring in a region), epidemic (an unusually high number of cases in a region), or pandemic (a global epidemic).

Transmission

An infectious disease is transmitted from some source. Defining the means of transmission plays an important part in understanding the biology of an infectious agent, and in addressing the disease it causes. Transmission may occur through several different mechanisms. Respiratory diseases and meningitis are commonly acquired by contact with aerosolized droplets, spread by sneezing, coughing, talking, kissing or even singing. Gastrointestinal diseases are often acquired by ingesting contaminated food and water. Sexually transmitted diseases are acquired through contact with bodily fluids, generally as a result of sexual activity. Some infectious agents may be spread as a result of contact with a contaminated, inanimate object (known as a fomite), such as a coin passed from one person to another, while other diseases penetrate the skin directly. Transmission of infectious diseases may also involve a vector. Vectors may be mechanical or biological. A mechanical vector picks up an infectious agent on the outside of its body and transmits it in a passive manner. An example of a mechanical vector is a housefly, which lands on cow dung, contaminating its appendages with bacteria from the feces, and then lands on food prior to consumption. The pathogen never enters the body of the fly.

In contrast, biological vectors harbour pathogens within their bodies and deliver pathogens to new hosts in an active manner, usually a bite. Biological vectors are often responsible for serious blood-borne diseases, such as malaria, viral encephalitis, Chagas disease, Lyme disease and African sleeping sickness. Biological vectors are usually, though not exclusively, arthropods, such as mosquitoes, ticks, fleas and lice. Vectors are often required in the life cycle of a pathogen. A common strategy used to control vector borne infectious diseases is to interrupt the life cycle of a pathogen by killing the vector.

The relationship between virulence and transmission is complex, and has important consequences for the long term evolution of a pathogen. Since it takes many generations for a microbe and a new host species to co-evolve, an emerging pathogen may hit its earliest victims especially hard. It is usually in the first wave of a new disease that death rates are highest. If a disease is rapidly fatal, the host may die before the microbe can get passed along to another host. However, this cost may be overwhelmed by the short term benefit of higher infectiousness if transmission is linked to virulence, as it is for instance in the case of cholera (the explosive diarrhea aids the bacterium in finding new hosts) or many respiratory infections (sneezing and coughing create infectious aerosols).

Preventing Transmission

One of the ways to prevent or slow down the transmission of infectious diseases is to recognize the different characteristics of various diseases. Some critical disease characteristics that should be evaluated include virulence, distance travelled by victims, and level of contagiousness. The human strains of Ebola virus, for example, incapacitate its victims extremely quickly and kills them soon after. As a result, the victims of this disease do not have the opportunity to travel very far from the initial infection zone.

Also, this virus must spread through skin lesions or permeable membranes such as the eye. Thus, the initial stage of Ebola is not very contagious since its victims experience only internal hemorrhaging. As a result of the above features, the spread of Ebola is very rapid and usually stays within a relatively confined geographical area.

In contrast, Human Immunodeficiency Virus (HIV) kills its victims very slowly by attacking their immune system. As a result, a lot of its victims transmit the virus to many others before even realizing that they are carrying the disease. Also, the relatively low virulence allows its victims to travel long distances, increasing the likelihood of an epidemic.

Another effective way to decrease the transmission rate of infectious diseases is to recognize the effects of small-world networks. In epidemics, there are often extensive interactions within hubs or groups of infected individuals and other interactions within discrete hubs of susceptible individuals. Despite the low interaction between discrete hubs, the disease can jump to and spread in a susceptible hub via a single or few interactions with an infected hub. Thus, infection rates in small-world networks can be reduced somewhat if interactions between individuals within infected hubs are eliminated. However, infection rates can be drastically reduced if the main focus is on the prevention of transmission jumps between hubs. The use of needle exchange programs in areas with a high density of drug users with HIV is an example of the successful implementation of this treatment method. Another example is the use of ring culling or vaccination of potentially susceptible livestock in adjacent farms to prevent the spread of the foot-and-mouth virus in 2001. General methods to prevent transmission of pathogens may include disinfection and pest control.

Diagnosis and Therapy

Diagnosis of infectious disease sometimes involves identifying an infectious agent either directly or indirectly. In practice most minor infectious diseases such as warts, cutaneous abscesses, respiratory system infections and diarrheal diseases are diagnosed by their clinical presentation. Conclusions about the cause of the disease are based upon the likelihood that a patient came in contact with a particular agent, the presence of a microbe in a community, and other epidemiological considerations. Given sufficient effort, all known infectious agents can be specifically identified. The benefits of identification, however, are often greatly outweighed by the cost, as often there is no specific treatment, the cause is obvious, or the outcome of an infection is benign. Specific identification of an infectious agent is usually only determined when such identification can aid in the treatment or prevention of the disease, or to advance knowledge of the course of an illness prior to the development of effective therapeutic or preventative measures.

For example, in the early 1980s, prior to the appearance of AZT for the treatment of AIDS, the course of the disease was closely followed by monitoring the composition of patient blood samples, even though the outcome would not offer the patient any further treatment options. In part, these studies on the appearance of HIV in specific communities permitted the advancement of hypotheses as to the route of transmission of the virus.

By understanding how the disease was transmitted, resources could be targeted to the communities at greatest risk in campaigns aimed at reducing the number of new infections. The specific serological diagnostic identification, and later genotypic or molecular identification, of HIV also enabled the development of hypotheses as to the temporal and geographical origins of the virus, as well as a myriad of other hypothesis. The development of molecular diagnostic tools have enabled physicians and researchers to monitor the efficacy of treatment with anti-retroviral drugs. Molecular diagnostics are now commonly used to identify HIV in healthy people long before the onset of illness and have been used to demonstrate the existence of people who are genetically resistant to HIV infection.

Thus, while there still is no cure for AIDS, there is great therapeutic and predictive benefit to identifying the virus and monitoring the virus levels within the blood of infected individuals, both for the patient and for the community at large.

Methods of Diagnosis

Diagnosis of infectious disease is nearly always initiated by medical history and physical examination. More detailed identification techniques involve the culture of infectious agents isolated from a patient. Culture allows identification of infectious organisms by examining their microscopic features, by detecting the presence of substances produced by pathogens, and by directly identifying an organism by its genotype.

Other techniques (such as X-rays, CAT scans, PET scans or NMR) are used to produce images of internal abnormalities resulting from the growth of an infectious agent. The images are useful in detection of, for example, a bone abscess or a spongiform encephalopathy produced by a prion.

Microbial Culture

Microbiological culture is a principal tool used to diagnose infectious disease. In a microbial culture, a growth medium is provided for a specific agent. A sample taken from potentially diseased tissue or fluid is then tested for the presence of an infectious agent able to grow within that medium. Most pathogenic bacteria are easily grown on nutrient agar, a form of solid medium that supplies carbohydrates and proteins necessary for growth of a bacterium, along with copious amounts of water. A single bacterium will grow into a visible mound on the surface of the plate called a colony, which may be separated from other colonies or melded together into a "lawn". The size, colour, shape and form of a colony is characteristic of the bacterial species, its specific genetic makeup (its strain), and the environment which supports its growth. Other ingredients are often added to the plate to aid in identification. Plates may contain substances that permit the growth of some bacteria and not others, or that change colour in response to certain bacteria and not others.

Bacteriological plates such as these are commonly used in the clinical identification of infectious bacteria. Microbial culture may also be used in the identification of viruses: the medium in this case being cells grown in culture that the virus can infect, and then alter or kill. In the case of viral identification, a region of dead cells results from viral growth, and is called a "plaque". Eukaryotic parasites may also be grown in culture as a means of identifying a particular agent.

In the absence of suitable plate culture techniques, some microbes require culture within live animals. Bacteria such as

Mycobacterium leprae and *T. pallidum* can be grown in animals, although serological and microscopic techniques make the use of live animals unnecessary.

Viruses are also usually identified using alternatives to growth in culture or animals. Some viruses may be grown in embryonated eggs. Another useful identification method is Xenodiagnosis, or the use of a vector to support the growth of an infectious agent.

Chaga's disease is the most significant example, because it is difficult to directly demonstrate the presence of the causative agent, *Trypanosoma cruzi* in a patient, which therefore makes it difficult to definitively make a diagnosis. In this case, xenodiagnosis involves the use of the vector of the Chaga's agent *T. cruzi*, an uninfected triatomine bug (subfamily Triatominae), which takes a blood meal from a person suspected of having been infected. The bug is later inspected for growth of *T. cruzi* within its gut.

Microscopy

Another principal tool in the diagnosis of infectious disease is microscopy. Virtually all of the culture techniques discussed above rely, at some point, on microscopic examination for definitive identification of the infectious agent. Microscopy may be carried out with simple instruments, such as the compound light microscope, or with instruments as complex as an electron microscope.

Samples obtained from patients may be viewed directly under the light microscope, and can often rapidly lead to identification. Microscopy is often also used in conjunction with biochemical staining techniques, and can be made exquisitely specific when used in combination with antibody based techniques.

For example, the use of antibodies made artificially fluorescent (fluorescently labelled antibodies) can be directed to bind to and identify a specific antigens present on a pathogen. A fluorescence microscope is then used to detect fluorescently labelled antibodies bound to internalized antigens within clinical samples or cultured cells. This technique is especially useful in the diagnosis of viral diseases, where the light microscope is incapable of identifying a virus directly.

Other microscopic procedures may also aid in identifying infectious agents. Almost all cells readily stain with a number of basic dyes due to the electrostatic attraction between negatively charged cellular molecules and the positive charge on the dye. A cell is normally transparent under a microscope, and using a stain increases the contrast of a cell with its background.

Staining a cell with a dye such as Giemsa stain or crystal violet allows a microscopist to describe its size, shape, internal and external components and its associations with other cells. The response of bacteria to different staining procedures is used in the taxonomic classification of microbes as well.

Two methods, the Gram stain and the acid-fast stain, are the standard approaches used to classify bacteria and to diagnosis of disease. The Gram stain identifies the bacterial groups Firmicutes and Actinobacteria, both of which contain many significant human pathogens. The acid-fast staining procedure identifies the Actinobacterial genera *Mycobacterium* and *Nocardia*.

Biochemical Tests

Biochemical tests used in the identification of infectious agents include the detection of metabolic or enzymatic products characteristic of a particular infectious agent. Since bacteria ferment carbohydrates in patterns characteristic of their genus and species, the detection of fermentation products is commonly used in bacterial identification. Acids, alcohols and gases are usually detected in these tests when bacteria are grown in selective liquid or solid media.

The isolation of enzymes from infected tissue can also provide the basis of a biochemical diagnosis of an infectious disease. For example, humans can make neither RNA replicases nor reverse transcriptase, and the presence of these enzymes are characteristic of specific types of viral infections. The ability of the viral protein hemagglutinin to bind red blood cells together into a detectable matrix may also be characterized as a biochemical test for viral infection, although strictly speaking hemagglutinin is not an *enzyme* and has no metabolic function. Serological methods are highly sensitive, specific and often extremely rapid tests used to identify microorganisms. These tests are based upon the ability of an antibody to bind specifically to an antigen. The antigen, usually a protein or carbohydrate made by an infectious agent, is bound by the antibody. This binding then sets off a chain of events that can be visibly obvious in various ways, dependent upon the test. For example, "Strep throat" is often diagnosed within minutes, and is based on the appearance of antigens made by the causative agent, *S. pyogenes*, that is retrieved from a patients throat with a cotton swab.

Serological tests, if available, are usually the preferred route of identification, however the tests are costly to develop and the reagents used in the test often require refrigeration. Some serological methods are extremely costly, although when commonly used, such as with the "strep test", they can be inexpensive. Complex serological techniques have been developed into what are known as Immunoassays. Immunoassays can use the basic antibody – antigen binding as the basis to produce an electro-magnetic or particle radiation signal, which can be detected by some form of instrumentation. Signal of unknowns can be compared to that of standards allowing quantitation of the target antigen.

To aid in the diagnosis of infectious diseases, immunoassays can detect or measure antigens from either infectious agents or proteins generated by an infected organism in response to a foreign agent. For example, immunoassay A may detect the presence of a surface protein from a virus particle. Immunoassay B on the other hand may detect or measure antibodies produced by an organism's immune system which are made to neutralize and allow the destruction of the virus. Instrumentation can be used to read extremely small signals created by secondary reactions linked to the antibody – antigen binding. Instrumentation can control sampling, reagent use, reaction times, signal detection, calculation of results, and data management to yield a cost effective automated process for diagnosis of infectious disease.

Molecular Diagnostics

Technologies based upon the polymerase chain reaction (PCR) method will become nearly ubiquitous gold standards of diagnostics of the near future, for several reasons. First, the catalogue of infectious agents has grown to the point that virtually all of the significant infectious agents of the human population have been identified. Second, an infectious agent must grow within the human body to cause disease; essentially it must amplify its own nucleic acids in order to cause a disease.

This amplification of nucleic acid in infected tissue offers an opportunity to detect the infectious agent by using PCR. Third, the essential tools for directing PCR, primers, are derived from the genomes of infectious agents, and with time those genomes will be known, if they are not already. Thus, the technological ability to detect any infectious agent rapidly and specifically are currently available.

The only remaining blockades to the use of PCR as a standard tool of diagnosis are in its cost and application, neither of which is insurmountable. The diagnosis of a few diseases will not benefit from the development of PCR methods, such as some of the clostridial diseases (tetanus and botulism).

These diseases are fundamentally biological poisonings by relatively small numbers of infectious bacteria that produce extremely potent neurotoxins. A significant proliferation of the infectious agent does not occur, this limits the ability of PCR to detect the presence of any bacteria.

Clearance and Immunity

Infection with most pathogens does not result in death of the host and the offending organism is ultimately cleared after the symptoms of the disease have waned. This process requires immune mechanisms to kill or inactivate the inoculum of the pathogen. Specific acquired immunity against infectious diseases may be mediated by antibodies and/or T lymphocytes. Immunity mediated by these two factors may be manifested by:

- a direct effect upon a pathogen, such as antibody-initiated complement-dependent bacteriolysis, opsonoization, phagocytosis and killing, as occurs for some bacteria,
- neutralization of viruses so that these organisms cannot enter cells,
- or by T lymphocytes which will kill a cell parasitized by a microorganism.

The immune system response to a microorganism often causes symptoms such as a high fever and inflammation, and has the potential to be more devastating than direct damage caused by a microbe. Resistance to infection (immunity) may be acquired following a disease, by asymptomatic carriage of the pathogen, by harboring an organism with a similar structure (crossreacting), or by vaccination. Knowledge of the protective antigens and specific acquired host immune factors is more complete for primary pathogens than for opportunistic pathogens. Immune resistance to an infectious disease requires a critical level of either antigen-specific antibodies and/or T cells when the host encounters the pathogen. Some individuals develop natural serum antibodies to the surface polysaccharides of some agents although they have had little or no contact with the agent, these natural antibodies confer specific protection to adults and are passively transmitted to newborns.

Historic Pandemics

A pandemic (or global epidemic) is a disease that affects people over an extensive geographical area.

- Plague of Justinian, from 541 to 750, killed between 50% and 60% of Europe's population.
- The Black Death of 1347 to 1352 killed 25 million in Europe over 5 years (estimated to be between 25 and 50% of the populations of Europe, Asia, and Africa-the world population at the time was 500 million).
- The introduction of smallpox, measles, and typhus to the areas of Central and South America by European explorers during the 15th and 16th centuries caused pandemics among the native inhabitants. Between 1518 and 1568 disease pandemics are said to have caused the population of Mexico to fall from 20 million to 3 million.
- The first European influenza epidemic occurred between 1556 and 1560, with an estimated mortality rate of 20%.
- Smallpox killed an estimated 60 million Europeans during the 18th century (approximately 400,000 per year). Up to 30% of those infected, including 80% of the

children under 5 years of age, died from the disease, and one third of the survivors went blind.

- In the 19th century, tuberculosis killed an estimated one-quarter of the adult population of Europe; by 1918 one in six deaths in France were still caused by TB.
- The Influenza Pandemic of 1918 (or the Spanish Flu) killed 25-50 million people (about 2% of world population of 1.7 billion). Today Influenza kills about 250,000 to 500,000 worldwide each year.

Emerging Diseases and Pandemics

In most cases, microorganisms live in harmony with their hosts. Such is the case for many tropical viruses and the insects, monkeys, or other animals in which they have lived and reproduced. Because the microbes and their hosts have co-evolved, the hosts gradually become resistant to the microorganisms. When a microbe jumps from a long-time animal host to a human being, it may cease to be a harmless parasite and become pathogenic. With most new infectious diseases, some human action is involved, changing the environment so that an existing microbe can take up residence in a new niche. When that happens, a pathogen that had been confined to a remote habitat appears in a new or wider region, or a microbe that had infected only animals suddenly begins to cause human disease. Several human activities have led to the emergence and spread of new diseases:

- Encroachment on wildlife habitats. The construction of new villages and housing developments in rural areas force animals to live in dense populations, creating opportunities for microbes to mutate and emerge.
- Changes in agriculture. The introduction of new crops attracts new crop pests and the microbes they carry to farming communities, exposing people to unfamiliar diseases.
- The destruction of rain forests. As countries make use of their rain forests, by building roads through forests and clearing areas for settlement or commercial ventures, people encounter insects and other animals harboring previously unknown microorganisms.

- Uncontrolled urbanization. The rapid growth of cities in many developing countries tends to concentrate large numbers of people into crowded areas with poor sanitation. These conditions foster transmission of contagious diseases.
- Modern transport. Ships and other cargo carriers often harbour unintended "passengers", that can spread diseases to faraway destinations. While with international jet-airplane travel, people infected with a disease can carry it to distant lands, or home to their families, before their first symptoms appear.
- Pollution of the environment. Changes in the climate (such as global warming) can cause microorganisms to adapt and create new strains, which can give them an evolution advantage.

The Study of Infectious Disease

History

Abû Alî ibn Sînâ (Avicenna) discovered the contagious nature of infectious diseases in the early 11th century. He introduced quarantine as a means of limiting the spread of contagious and infectious diseases in *The Canon of Medicine*, *circa* 1020. He also stated that bodily secretion is contaminated by foul foreign earthly bodies before being infected, but he did not view them as primary causes of disease.

When the Black Death bubonic plague reached al-Andalus in the 14th century, Ibn Khatima and Ibn al-Khatib hypothesized that infectious diseases are caused by "contagious entities" which enter the human body. Such ideas became more popular in Europe during the Renaissance, particularly through the writing of the Italian monk Girolamo Fracastoro. Anton van Leeuwenhoek (1632-1723) advanced the science of microscopy by being the first to observe microorganisms, allowing for easy visualization of bacteria. Louis Pasteur proved beyond doubt that certain diseases are caused by infectious agents, and developed a vaccine for rabies. Robert Koch, provided the study of infectious diseases with a scientific basis known as Koch's postulates. Edward Jenner, Jonas Salk and Albert Sabin developed effective vaccines for smallpox and polio, which would later result in the eradication and near-eradication of these diseases, respectively.

Alexander Fleming discovered the world's first antibiotic Penicillin which Florey and Chain then developed. Gerhard Domagk developed Sulphonamides, the first broad spectrum synthetic antibacterial drugs.

Medical Specialists

The medical treatment of infectious diseases falls into the medical field of Infectiology and in some cases the study of propagation pertains to the field of Epidemiology. Generally, infections are initially diagnosed by primary care physicians or internal medicine specialists. For example, an "uncomplicated" pneumonia will generally be treated by the internist or the pulmonologist (lung physician). The work of the infectiologist therefore entails working with both patients and general practitioners, as well as laboratory scientists, immunologists, bacteriologists and other specialists..

An infectious disease team may be alerted when:

- The disease has not been definitively diagnosed after an initial workup
- The patient is immunocompromised (for example, in AIDS or after chemotherapy);
- The infectious agent is of an uncommon nature (e.g. tropical diseases);
- The disease has not responded to first line antibiotics;
- The disease might be dangerous to other patients, and the patient might have to be isolated

International Committee on Taxonomy of Viruses and the 3,142 Unassigned Species

The VIIIth ICTV Report lists more than 6,000 viruses classified in 1,950 species, themselves classified in 391 taxa. Thus, it was surprising to discover that GenBank (part of the International Nucleotide Sequence Database Collaboration, comprising the DNA DataBank of Japan, the European Molecular Biology Laboratory, and GenBank, which is located at the National Centre for Biotechnology Information (NCBI) in the U.S.

National Institutes of Health) has collected sequences belonging to 3,142 "species" of viruses not present in the ICTV current master list of 2005.

For the past 20 years, GenBank has increased the number of sequences stored from one in 1982 to 42 million in 2002, and has increased the number of tools at the disposal of users to the point where today it is an absolutely necessary system for molecular biologists and virologists worldwide.

In particular, GenBank has developed a complete database for taxonomy, including virus taxonomy, allowing virologists to search all viruses and to identify new sequences and new virus names. Were it not for the existence of this excellent system, we would not have known of the existence of these 3,142 unassigned "species" of viruses.

ICTV also has been very active in the last 20 years, incrementally increasing the number of taxa and virus names from 369 in 1985 to 7,881 in 2004. Not only have the numbers increased exponentially (about 20-fold), but the complexity of virus nomenclature and taxonomy has become tremendously complicated and controversial, even puzzling for some.

However, the overall stability of this virus classification, established in 1962, is quite remarkable in that, for example, names of all genera and families established in the 1980s are still in use in 2005. The advancement with the most impact was the definition of a virus species, which still is not fully understood by most virologists. Therefore, it is not surprising that its application in terms of names and concept causes problems such as the one discussed in this chapter.

The listing by GenBank, but not by ICTV, of 3,142 unassigned "species" clearly demonstrates a general problem of the application of the definition of virus species and that ICTV and GenBank must work in concert to cope with the day-to-day reality of virology and virus classification, and that collectively we need to improve the system in such a way that we will rapidly classify so-called "species" of viruses and change the current system, so that we do not find ourselves in a similar situation in the future.

This chapter will review the probable causes of the present situation and will identify actions that could remedy the situation both at GenBank and at ICTV. Finally a brief description of the Taxonomic Proposals Management System (TPMS), a new database to electronically handle all taxonomic proposals for ICTV, will be provided.

Review of the Species Concept

A virus species, as with all other taxonomic levels, is a concept devised by humans to describe concrete (real) biological viral entities. The definition of a virus "species", established and accepted by ICTV, is as follows: "A virus species is a polythetic class of viruses that constitutes a replicating lineage and occupies a particular ecological niche".

This definition is broad enough to allow considerable flexibility. It is the responsibility of the virologists interested in each virus family to apply this definition to their favourite viruses and to decide how many and which species should comprise a particular genus of a particular family. Typically, this is the work of ICTV study-groups, whose members are experts in particular virus families. Virus sequences are physical realities that are not species but entities assigned to a particular species. For example, the 24 bluetongue viruses are assigned to the species *Bluetongue virus* in the genus *Orbivirus*, family *Reoviridae*; thus, 24 viruses assigned to a single species. Therefore sequences entered in GenBank must be associated with the name of a virus isolate.

The term virus isolate has no specific definition and can be applied to any virus, so long as that virus has existed for some time. Several terms are used by virologists to define a virus entity below the species level and are somewhat variable and confusing, but for virus sequences the term "virus isolate" seems the most appropriate, practical and useful.

Sequence and Information Collection

The intent of GenBank is to collect sequences but the author, the person who enters the sequences in the database, is responsible for the accuracy of the sequences as well as for the validity of the associated information about the sequences, which includes the taxonomic information about the virus in question. The primary taxonomic information pertains to an isolate of a previously recognized virus or of a newly recognized virus. Consequently, if the author errs in naming the virus or in its classification, the entry will be incorrect and might remain incorrect for a long period of time.

GenBank has personnel qualified to review all entries and they effectively and immediately correct a large number of obvious mistakes. However, with regard to "new species", GenBank personnel do not have the means to determine whether this new taxon is justified and acceptable or not. Obviously, this policy of GenBank cannot be changed from without, but it is certainly possible for GenBank to slightly modify the Entrez form in order to build in safety nets to avoid such situations in the future.

GenBank Entrez form and Software

The Entrez form of GenBank is a general form for all organisms, and one of the first questions asks the name of the organism. Whereas viruses are not organisms, this question immediately causes ambiguity because it is not clear whether the isolate name or the name of the species to which it belongs should be indicated, notwithstanding the fact that many investigators have not yet understood the difference between an isolate name, corresponding to a specific sequence of a virus, and the conceptual species name, to which a particular isolate has been assigned.

The second drawback comes from the fact that the software dealing with Entrez and connected to the Taxonomy database of GenBank will by default assign any new virus name in the field "Organism" to a new "species" if it does not exist in the ICTV Master list, from which the current valid classification is derived.

This implies that most of the 3,142 "species" are in fact virus isolates only (concrete entities), rather than species (conceptual). The third problem arises from the fact that many virus species names and virus names are identical in the words that compose them, except that species names are written in italics and are not abbreviated, while virus names are not italicized and may be abbreviated; unfortunately, databases can only use ASCII characters and cannot accept italics.

It is particularly unfortunate that we are in this situation and that different names have not been created for virus species, but this is a reality that cannot be changed, at least in the short term, and consequently we need to find a way to finesse these problems and still adhere to acceptable virus nomenclature.

It seems that the only possibility is to redesign software to show taxonomy as it should be shown, and to restore the difference between the two names, if not by italics at least by some equivalent means. The fourth problem also is associated with the ASCII code and that is the impossibility of using Greek letters, which are incorporated in many phage names.

For this, one solution might be to have ICTV change these names slightly by spelling out the Greek letters, c.f., "Enterobacteria phage lambda", instead of "Enterobacteria phage e".

Exponential Increase in Sequencing

Because of lower costs and more efficient technology, sequencing is increasing exponentially. Known viral sequences already are in the hundreds of thousands and it is conceivable that millions more will become available in the coming years. The number of virus taxa and isolates also increases exponentially and is therefore linearly related to sequencing activities. It is therefore critical to have all data stored and named properly and with taxonomic updates in real time.

Taxonomic Knowledge and Consciousness of Virologists

Not all virologists determining sequences are fully aware of or interested in virus taxonomy. These people are the primary source of information and they consequently are the primary sources of data entry errors. Left adrift, they will not make efforts to solve these particular sorts of problems, so other solutions should be put in place to remedy the problem. Insisting that they follow taxonomic rules seems incompatible with the general "author" principle of GènBank, but it is possible to "invite" them to enter the proper information directly on the Entrez form.

ICTV Organization

The organization structure of ICTV, the generally accepted leader in viral taxonomy, can be conceived of as pyramidal. The base of the pyramid comprises more than 500 virologists in six Subcommittees, divided into 82 specialty study-groups (SG), each representing a family of viruses or an unassigned genus. Each SG comprises from three to 15 experts selected for their expertise in recognized and new viruses and new sequences of viruses in each family; obviously this is an increasingly difficult task, given the rapid increase in available data. These 82 SGs present taxonomic proposals from the level of species to the level of order, the proposals are reviewed by the Executive Committee (EC) of the ICTV and, 18 months later, through a posting on ICTVnet and two rounds of discussions, are then ratified (or not) and voted upon (or not) by the ICTV membership at large.

This system, although well organized and in full compliance with the international code for virus taxonomy and nomenclature, is neither adapted to a large number of new isolates and species nor to a quick acceptance, so that changes are envisioned. In addition, by statute, the ICTV does not have the mandate to deal with virus isolates, which are not taxa.

This is a fundamental problem because no organization is responsible for certifying isolate names or for determining to which virus species they belong. However, ICTV, for historical reasons, lists a large number of unclassified virus isolates and SG members are the most knowledgeable people to deal with such an issue. Obviously, this should be considered by ICTV in order to solve this issue of unassigned viruses erroneously purported to be "species" and found in various databases.

What are the Solutions?

It is not the responsibility of the authors to decide whether their new virus isolates belong to an existing or a new species, not the responsibility of GenBank to police authors and organize virus taxonomy and nomenclature, and not the responsibility of ICTV to deal with non-taxonomic issues, such as isolate names, but it is our communal responsibility to work together to make the system functional; and it certainly is feasible.

Here we call for collaboration of all parties to establish a generally satisfactory system for virus taxonomy and nomenclature of virus isolates, sequence data, and all other virological data. Obtaining solutions will necessitate the collaboration of all those involved in producing, collecting and classifying viruses.

Classifying the 3,142 Unassigned Virus Names in GenBank

Placing the 3,142 unassigned "species" (which likely are simply "viruses") registered at GenBank will require manual classification by the ICTV SG members. Most of these "species" are probably isolates of recognized species and are in the GenBank database by default as new "species" because, whether they have been named or not, they have not been entered in the GenBank forms.

Some of these GenBank "species", however, may actually be new species that have not yet been submitted to ICTV and, if so, they should be proposed through the normal channels of the Taxonomic Proposals (TaxoProps) for evaluation and official ratification before they can be recognized as new species.

Improving the Management of Taxonomic Proposals

As a remedy to the general problem of virus classification, we suggest a two-tier plan. Firstly, GenBank would work with ICTV to provide to the authors rolling menus, or automatic writing, offering a list of all the official ICTV names to select from. This would help avoid spelling errors or the creation of incorrect new names by authors.

In the case of an unaccepted (or provisional) virus name, the default should be "unassigned virus in the genus, or in the family, or in the order or in the virus kingdom". This assignation to an "unassigned category" would trigger an email to the relevant ICTV SG chair and to some EC members so that ICTV will be immediately aware of this new entry. This information also would be stored in a proposed new database called TaxoProp Management System (TPMS), with a pending status to be resolved promptly. The second tier of the remedy would be to establish the TPMS database to handle all new TaxoProps containing information originating from an SG member, from any individual virologist or directly from GenBank.

Changing ICTV to Work in Real Time

Resolving the mechanics of the system is only a partial solution and more details are needed in order for this to be functional in real time. ICTV has established a sophisticated and democratic system to receive, review, discuss, re-discuss and finally approve taxonomic proposals, whether they concern a new species or a new order. This system is very sophisticated in principle but is too complex and takes too long to be compatible with modern virology work. There are, however, fundamental and practical differences between a TaxoProp for a new species and a TaxoProp for everything else, and that is that species are solely determined by specialty groups (ICTV SGs) and rarely by the public or the ICTV EC members.

The only envisioned ICTV concern would be to make certain that the virus species name follows the ICTV International Rules. It would therefore be possible to computerize the verification of the names in order to "bypass" ICTV EC scrutiny and only then obtain approval of the ICTV membership by electronic voting.

This in turn would considerably reduce the turn-around time: essentially, as soon as a name would be proposed by an ICTV SG, it would be automatically verified and accepted if all the rules were followed.

Increasing Relationships Between ICTV and GenBank

In order to have taxonomic consistency, it is absolutely essential that we strengthen the relationships between ICTV and GenBank. This could be done with an electronic link between TPMS and GenBank, supplemented with improved and regular personal relations.

For example, GenBank has appointed virus experts for a number of families of viruses. It would be more efficient to have GenBank somehow linked directly to the SG chairs of ICTV, so that GenBank would get taxonomic experts for advice in virus taxonomy.

This, in turn, would provide a more active role for the SG members and would activate the overall system of virus taxonomy.

The TPMS Database?

Due to the rapid increase in the number of TaxoProps and the increasing difficulty in manually handling several thousand taxonomic names, it is planned by ICTV to create a database to handle all ICTV taxa, from inception to acceptance, and from species to order.

Overall Description of the TPMS

The goal of the TPMS database is to electronically handle all the virus taxa at large (including virus isolates), i.e., those that have been inherited from previous reports, those that have been approved between 1999 and 2005, those in the process of being approved and those that will be approved in the future.

At any time it will then be possible to:

1- certify the approval status of any TaxoProp

2- elaborate a complete list of approved taxa

3- establish an electronic link with GenBank and all other virus databases.

Structure of the TPMS Database

The database will be accessible on line to all ICTV members and to the public at large. The database will comprise a collection of individual files, each one corresponding to a single taxon (order, family, subfamily, genus, and species), and in addition will record isolates. Each TaxoProp will progress from the initial proposal to the approval of the taxon by the ICTV. The bulk of the database will be constituted by all the approved taxa and individual viruses and should be exportable to other databases for future use.

Approval Process of the TaxoProps

The TaxoProp approval system of ICTV. When a TaxoProp form is filed on line, it is then considered by ICTV and becomes "Considered". A TaxoProp will consist of the proposal itself, of arguments provided to support the TaxoProp, and by attached documents to be downloaded by the reader. A unique TaxoProp number will be automatically generated when a form is completed or when a TaxoProp is modified by the author.

After consideration, it becomes "Considered", and comments of the ICTV are added on line and available to the public at large. Comments can be added by any virologist and the author of the TaxoProp can respond on line. The change of level will be done through a voting system in which ICTV EC members can view the votes at any time and will be operated by the TPMS curator. When accepted by ICTV, the TaxoProp becomes "Accepted" and comments are no longer added. When the ICTV at large has voted to approve the TaxoProp, it becomes "Voted" and is *de facto* a ratified taxon of the ICTV.

Filing TaxoProps

TaxoProps can be filed by any virologist, but it is likely that most will be filed by SG members. There will be a set of TaxoProps to choose from on line according to the sort of taxon proposed: an isolate (although not a formal taxon), a species, a type species, a genus, a subfamily, a family, an order, or any other taxon proposed by any virologist. Higher taxa cannot be proposed unless a lower level has been proposed and accepted. A genus must have a type species and at least one species or a list of species in the genus. Any proposed new taxa must have a name following the ICTV Code.

Link with Species Demarcation Criteria Lists

For each genus, it will be possible to list the species demarcation criteria of that genus for the benefit of the author and readers of the species TaxoProps. These lists will be extracted from the VIIIth Report and from subsequent updates. It is envisioned that these lists of species demarcation criteria will become useful tools for ICTV to search for a better means of making the overall virus taxonomy more uniform in the future.

Special Case of TaxoProp Isolates

Although isolates do not comprise formal taxa, recording and listing them are indispensable for the creation of species taxa and also for practical purposes. Virus species being defined from properties of virus isolates, ICTV records the existence of isolates in order to define species. There are no formal rules for naming isolates and, consequently, any name could potentially be accepted. However, it is common sense that these names follow previously established names and follow somewhat virus species names rules to avoid complete chaos.

Accepting TaxoProp Species

TaxoProp species, after being proposed by the SGs, will automatically be reviewed to ensure that they follow the ICTV International Code and, if they do, will automatically be approved electronically. However, a vote by the ICTV membership at large will still be necessary for it to be considered as ratified by the ICTV; this also could be done electronically.

Approved Taxa

All approved taxa, because they have been inherited, will be entered in the TPMS, so that the database will be updated automatically and in real time and can be linked to GenBank, ICTVdb (ICTV database) and other databases. Files for each taxon will archive all the comments and votes so that we can always return to this information if needed. However, public access to the TPMS for the ratified ICTV Master List will remain under control of the President of ICTV.

Generating the Ratified list of Taxa

Using TPMS, it will be possible to generate a complete list of approved taxa at any time, and this will be possible either alphabetically or taxonomically. Accepted taxa also will be presented according to the order of presentation of virus taxonomy: dsDNA, ssDNA, rtDNA, rtRNA, dsRNA, ssNRNA, ssPRNA, SAT (Satellites), VIR (Viroids), UN (unassigned). Within each category, the families will be classified according to the Order of Presentation of the Virus. Taxonomy Species will be assigned to a genus and classified alphabetically within the genus. The species taxa should comprise at least one isolate but include as many as described.

Searching the TPMS Database

Any virologist will have access to the TPMS database on line anytime so that he or she can verify names, either with an alphabetical pull down menu system or by use of a search engine.

Linking the TPMS Database to other Databases

TPMS will *de facto* become the definitive reference for virus taxonomy and nomenclature. Once approved by the ICTV President, typically after each ICTV EC meeting or after a certain period of time (such as six months), the TPMS database will be electronically linked with GenBank, ICTVdb, and other associated databases to ensure complete homogeneity of the virus taxonomy and nomenclature. This tool will be particularly important for the members of the International Virus Database

Network, who recognize the critical need for such a system, and to homogenize and update in real time the virus taxonomy and nomenclature.

Conclusion

Because sequencing is much less expensive and increasingly done these days, it is assumed that investigators will generate many more sequences of known and unknown viruses in the very near future.

The number of sequences is increasing exponentially and so will the number of virus names and virus taxa, as they are directly correlated with sequencing activities. It is clear that past management of virus taxonomy will become inadequate commensurate with this increase and that more rapid and better-adapted management is required. The electronic connection between databases storing sequences, storing biological data or handling taxonomic proposals is the only solution for the future. Human intervention is indispensable, but should be restricted to the decision-making process; all other interventions might best be done electronically.

This will mean better collaboration between the people responsible for the different types of databases, the people involved in the ICTV organization, and individual investigators themselves. Software will have to be adapted or devised to better fit virus taxonomy and nomenclature, ICTV will have to better appreciate day-to-day realities such as dealing with virus isolates, and scientists will have to be better educated in virus taxonomy. Regarding the decision-making process for virus species and isolates, a much greater role for the ICTV SGs will be necessary because the ultimate expertise resides with them in determining which new viruses fit into existing taxa and which must be placed in defined new taxa.

Consequently, it would be best for ICTV to work more closely with GenBank to establish the electronic TPMS database to work in real time and to cope with an acutely burgeoning number of taxa.

International Committee on Taxonomy of Viruses

The International Committee on Taxonomy of Viruses (ICTV) is a committee which authorizes and organizes the taxonomic classification of viruses. They have developed a universal taxonomic scheme for viruses and aim to describe all the viruses of living organisms. Members of the committee are considered to be world experts on viruses. The committee formed from and is governed by the Virology Division of the International Union of Microbiological Societies. The committee also operates an authoritative database (ICTVdB) containing taxonomic information for over 1,500 virus species, as of 1999. It is open to the public and is searchable by several different means.

Objectives

The official objectives of the ICTV are:

1. To develop an internationally agreed taxonomy for viruses
2. To develop internationally agreed names for virus taxa, including species and subviral agents
3. To communicate taxonomic decisions to all users of virus names, in particular the international community of virologists, by publications and via the Internet
4. To maintain an index of virus names
5. To maintain an ICTV database on the Internet, that records the data that characterize each named viral taxon, together with the common names of each taxon in all major languages.

Principles of Nomenclature

The ICTV's essential principles of virus nomenclature are:

- Stability
- To avoid or reject the use of names which might cause error or confusion
- To avoid the unnecessary creation of names.

The ICTV's universal virus classification system uses a slightly modified version of the standard biological classification system. It only recognises the taxa order, family, subfamily, genus, and species. When it is uncertain how to classify a species into a genus but its classification in a family is clear, it will be classified as an unassigned species of that family. Many taxa remain unranked.

Naming and Changing Taxa

Proposals for new names, name changes, and the establishment and taxonomic placement of taxa are handled by the Executive Committee of the ICTV in the form of proposals. All relevant ICTV subcommittees and study groups are consulted prior to a decision being taken.

The name of a taxon has no status until it has been approved by ICTV, and names will only be accepted if they are linked to approved hierarchical taxa. If no suitable name is proposed for a taxon, the taxon may be approved and the name be left undecided until the adoption of an acceptable international name, when one is proposed to and accepted by ICTV. Names must not convey a meaning for the taxon which would seem to either exclude viruses which are rightfully members of that taxa, exclude members which might one day belong to that taxa, or include viruses which are members of different taxa.

Rules for Taxa

Species

A species name shall consist of as few words as practicable but must not consist only of a host name and the word *virus*. A species name must provide an appropriately unambiguous identification of the species. Numbers, letters, or combinations thereof may be used as species epithets where such numbers and letters are already widely used. However, newly designated serial numbers, letters or combinations thereof are not acceptable alone as species epithets. If a number or letter series is in existence it may be continued.

Genera

A virus genus is a group of related species that share some significant properties and often only differ in host range and virulence. A genus name must be a single word ending in *virus*. Approval of a new genus must be accompanied by the approval of a type species.

Subfamilies

A subfamily is a group of genera sharing certain common characters. The taxon shall be used only when it is needed to solve a complex hierarchical problem. A subfamily name must be a single word ending in *virinae.*

Families

A family is a group of genera, whether or not these are organized into subfamilies, sharing certain common characters. A family name must be a single word ending in *viridae.*

Orders

An order is a group of families sharing certain common characters. An order name must be a single word ending in *virales.*

Rules for Sub-viral Agents

Rules concerned with the classification of viruses shall also apply to the classification of viroids. The formal endings for taxa of viroids are the word *viroid* for species, the suffix-*viroid* for genera, the suffix-*viroinae* for sub-families, should this taxon be needed, and-*viroidae* for families. Retrotransposons are considered to be viruses in classification and nomenclature.

Satellites and prions are not classified as viruses but are assigned an arbitrary classification as seems useful to workers in the particular fields.

Rules for Orthography

1. In formal taxonomic usage the accepted names of virus orders, families, subfamilies, and genera are printed in italics and the first letters of the names are capitalized.
2. Species names are printed in italics and have the first letter of the first word capitalized. Other words are not capitalized unless they are proper nouns, or parts of proper nouns.
3. In formal usage, the name of the taxon shall precede the term for the taxonomic unit.

Database

The development of ICTVdB has been supported by the ICTV since 1991 and was initially intended to aid taxonomic research. The database classifies viruses based primarily on their chemical characteristics, genomic type, nucleic acid replication, diseases, vectors, and geographical distribution, among other characteristics. The database was developed at the Australian National University with support of the US National Science Foundation, and sponsored by the American Type Culture Collection.

It uses the Description Language for Taxonomy (DELTA) system, a world standard for taxonomic data exchange, developed at Australia's Commonwealth Scientific and Industrial Research Organisation (CSIRO). DELTA is able to store a wide diversity of data and translate it into a language suitable for traditional reports and web publication. For example, ICTVdB does not itself contain genomic sequence information but can convert DELTA data into NEXUS format. It can also handle large data inputs and is suited to compiling long lists of virus properties, text comments, and images.

ICTVdB has grown in concept and capability to become a major reference resource and research tool, in 1999 it was receiving over 30,000 combined online hits per day from its main site at the Australian National University, and two mirror sites based in the UK and USA.

Viral Diseases of Poultry: The Tip of the Iceberg

If you ask poultry producers what damage viruses can do to their business, they will understandably think mostly about major diseases such as avian influenza, Gumboro disease or respiratory infections such as Infectious Bronchitis virus. However, this is merely the tip of the iceberg. Viral challenges can be the triggers for a whole range of problems which may never even appear as clinical disease but can have an even more devastating effect on overall flock performance, production and profits.

Poultry disease can affect one or more of the birds' body systems, the most significant of which are the respiratory tract (lungs and air sacs), the skeletal system (bones and joints) or the intestinal tract (affecting gut function).

Although other factors such as environment, nutrition and management play an important role in the full expression of diseases affecting these systems, the most significant underlying trigger is usually an infectious agent, and the most potent of these are undoubtedly viruses.

Viruses do their damage by causing primary tissue damage or by opening the gates for other infectious agents, such as bacteria and mycoplasmas, which might also be present, just waiting for a chance to act. And if that was not enough, a number of avian viruses can take this insidious yet highly damaging effect up to another level by directly attacking the bird's immune system.

This can cause clinical problems in its own right or prevent the bird from recognising other harmful agents. It can also prevent birds from responding to the increasing number of vaccinations given to a range of birds.

The positive benefits of the effective control of virus diseases have been known for over 200 years since Edward Jenner's pioneering work on smallpox vaccination. In the intervening period, these concepts have been extended to animal populations worldwide. However, even in the 21st century, viral disease has continued to be a limiting factor to full expression of optimal performance in livestock species despite the emergence of large numbers of high quality vaccines.

Why are Viruses Potentially More Important than Bacteria and Other Infectious Agents?

The first is their frequency. Common things are common. Poultry species can be exposed to a wide variety of viruses, even on high health status farms. Viruses can persist in the farm environment and diagnostic tests can be laborious, expensive and may make detection difficult.

The next important factor is their impact. Primary viral infections can open the way for other secondary bacterial infections to cause on-going chronic and severe performance losses. Viruses are frequently sub-clinical and can be part of complex syndromes that are difficult to unravel and hence manage. Finally, all these aspects make viruses difficult to control. Being sub-clinical and undetected, they can spread easily. They are not treatable with antibiotics.

Viruses can therefore build up on farms especially if they are multi age and this can put too large a burden on vaccines that are considered by some to be the only effective defence against viruses. However, these vaccines can only work to their full potential if the majority of birds in a flock have an immune system that can respond to the vaccine and produce a robust and long lasting immunity.

All of this is not just scientific theory, it is economic fact. To illustrate these points, it is worth considering some examples of the effects of viruses:

Gut disease, malabsorption and stunting. One of the most economically significant diseases of the turkey industry in recent years has been the Poult Enteritis Mortality Syndrome (PEMS).

This disease can cause mortality rates up to 80 per cent and severe clinical stunting problems, poor growth and high levels of culling of birds that cannot reach market weight. Some flocks have been totally destroyed by this syndrome.

Research has shown a variety of viruses to be involved and the most significant candidates are coronaviruses, rotaviruses and astroviruses.

Control of these viruses by terminal cleansing and disinfection using a virucidal programme has been shown to be the only truly effective measure against this syndrome.

As an example, research at the South Eastern Poultry Research Laboratory, Athens, Georgia, discussed the importance of Virkon® S (Dupont Animal Health Solutions) in inactivating astroviruses. Virkon S was shown to be superior to other products available.

Respiratory Disease

Respiratory disease syndromes are usually complex interactions of an assortment of infectious agents. Classically, Chronic Respiratory Disease (CRD) in broilers was associated with infectious bronchitis virus, Newcastle disease virus, mycoplasma and *E. coli* working in tandem.

In turkeys, avian pneumovirus (turkey rhinotracheitis virus), Newcastle disease virus, mycoplasmas, *Ornithobacter rhinotrachealae* (ORT) and *E. coli* are involved in a number of respiratory complexes.

Recent research has shown the importance of viral triggers in making the effects of the secondary bacterial infections more severe and long lasting by allowing them to 'leap-frog' over the birds' defence mechanisms.

Naylor et al. (1992) showed that a combination of *Mycoplasma gallisepticum* and avian pneumovirus produced more severe clinical signs in infected turkey poults and increased the length of time over which clinical disease occurred than for each agent individually.

Similarly, Marien et al. (2005) showed that the bacterial infection ORT alone was unable to produce clinical disease in experimental turkeys but when combined with avian pneumovirus, produced significant respiratory disease, which was more long lasting than avian pneumovirus infection alone.

An Assault on the Immune System

It has been known for many years that Gumboro disease (infectious bursal disease, IBD) virus, chick anaemia virus and Marek's disease virus can have a direct effect on the major tissues of the bird involved in producing immune defences, especially the bursa, thymus and spleen.

Indeed there is evidence that combinations of two or more of these viruses can enhance their effects on birds. In turkeys, haemorrhagic enteritis virus (HEV) plays a similar role. Marek's disease virus causes direct and indirect economic losses through this damage to the immune system. There may be direct mortality or increased susceptibility to common bugs such as *E. coli.*

Gumboro disease virus can result in transient or long lasting damage to tissues of the bursa of Fabricius, the organ of the bird that is the cornerstone in producing antibodies and defence against disease. Perhaps the clearest illustration of economic effects of sub-clinical disease associated with viral infection is in relation to chick anaemia virus (CAV) infection.

Research at the Veterinary Research Laboratory, Stormont, Belfast, Northern Ireland confirmed that broiler flocks affected by clinical CAV infection from a vertically transmitting parent flock had a net income per 1000 birds 17 to 19% lower, average slaughter weight 3.3% lower and mortality rate 2% higher than that recorded in unaffected flocks.

Other work suggests that the depression of average weight can be as much as 12.8%. Even more significant is the sub-clinical effects of CAV infection.

Workers at the Veterinary Research Laboratory, Stormont in Northern Ireland, demonstrated on a large-scale trial looking at the results of 1,000,000 broilers that horizontally acquired CAV infection, i.e. persistent virus on the broiler farm, was capable of causing significant economic loss resulting from sub-clinical infection. Even in flocks not reporting any obvious disease signs, the infection was responsible for:

1. 13% drop in net income per 1000 birds
2. 2% penalty in feed conversion ratio (FCR)
3. 2.5% depression in average live weight at slaughter.

Another demonstration of the severe economic effect of immuno-suppression is thought to be the trigger for the largest cause of condemnation of broilers in the USA, namely infectious process, IP, a form of avian cellulitis.

This subcutaneous bacterial infection appears to gain access through skin wounds and it is considered that birds with a poorly functioning immune system are much more prone to this condition. In the USA alone, losses of $40-50 million per annum are conservatively estimated due to avian cellulitis.

Where do We go from Here?

It is easy to demonstrate the dramatic clinical effect of viral infections on poultry flocks and there is a wealth of science confirming the insidious day in, day out effects of viral infections on bird performance without them even showing any clinical disease.

The variety of viral agents involved coupled with their frequency and persistence means that controlling viruses must be the foundation for any disease control biosecurity programme for poultry farms. Although there is an armoury of vaccines available, it is clear that for them to work most effectively they should be given the optimal opportunity to fulfil their potential.

Reducing the pressure on vaccines by reducing the viral load is the way to realise that potential. Effective measures are available with attention to structured and targeted cleansing and disinfection using disinfectant products with a proven virucidal pedigree.

In a number of livestock species, Virkon S has shown itself to be the thoroughbred in virus control with a broad spectrum of activity (including against the secondary bacteria, mycoplasma, moulds and fungi), maintaining that activity in the presence of organic challenge.

Biochemical Test for Viability

What is Seed Viability?

The viability of the seed accession is a measure of how many seeds are alive and could develop into plants which will reproduce themselves, given the appropriate conditions.

Why do we Test Seed Viability?

It is important to know that the seeds that are stored in a genebank will grow to produce plants. Therefore they must have a high viability at the start and during storage. The viability of seeds at the start of storage will also determine, within the environmental conditions, the storage life of the accession.

When should Viability be Determined?

Viability will need to be determined at the start of storage and at regular intervals during storage to predict the correct time for regeneration of the accession.

The viability test takes from a few days to weeks or even months to give an accurate result. If possible the results should be available before the seeds are packaged and placed in the genebank so that poor quality seeds can be identified and regenerated before storage.

Where the viability cannot be determined before storage, the seeds should be placed into long-term storage to ensure their safety whilst awaiting the results of the test.

How should Viability be Determined?

The most accurate test of viability is the germination test and this will be described here. The germination test is made under controlled conditions to find out how many seeds will germinate and produce normal seedlings which could develop into normal reproductively mature plants.

The IBPGR Advisory Committee on Seed Storage recommends that for the initial germination test of species where a reasonable germination technique is available, a minimum of two replicates using 200 seeds (100 seeds per replicate) is acceptable, providing that germination is above 90%.

If not, a further 200 seeds should be tested as before and the overall result for seed viability taken as the mean of the two tests. Other biochemical tests are available to test viability.

These have the advantage of being quicker, but are not as accurate and require considerable skill and practice in their implementation and interpretation. These are not recommended by the IBPGR Advisory Committee on Seed Storage for general use as tests for seed viability.

Bibliography

Alfred Byrd Graf: *Advances in Plant Physiology*, Rajat Pub, Delhi, 2008.

Alfred Steferud: *Diseases of Fruits and Nuts*, Biotech Books, Delhi, 2005.

Bag, A. K.: *Mathematics in Ancient and Medieval India*, Varanasi: Chaukhambha Orientalia, 1979.

Bahar A. Siddiqui and Samiullah Khan: *Plant Breeding Advances and in vitro Culture*, CBS, Delhi, 1997.

Bandyopadhyay, P C : *Breeding and Crop Production*, Gene Tech Books, Delhi, 2007.

Collymore L.: *Fruit Production in Barbados*, Port of Spain, Trinidad and Tobago, 1996.

Daniells J.: *Illustrated G* [illegible] *ation of Banana varieties in the South Pacifi*[illegible], [illegible] R, 1995.

Fageria, N.K.; V C Baligar and R B Clark: *Physiology of Crop Production*, International Book, Delhi, 2007.

Fortuner, R.: *Nematode Identification and Expert System Technology,* New York, Plenum Press, 1988.

Fowler, David: *The Mathematics of Plato's Academy*, Oxford, Clarendon, 1999.

Georges, S.: *The Debt Boomerang: How Third World Debt Harms Us All*, Boulder, Westview Press, 1992.

Hardy B.: *Biology and Agronomy of Forage Arachis*, Cali, International Centre for Tropical Agriculture, 1994.

Jones, R. M.: *Plant Resources of South-East Asia,* Wageningen, Pudoc Scientific Publishers, 1992.

Julia F.: *Fruits of Warm Climates*, Miami, Julia F. Morton Publisher, 1987.

Kapoor, R.L. and M.L. Saini: *Plant Breeding and Crop Improvement*, CBS, Delhi, 1997.

Mishra, Shubhrata R. : *Morphology of Fungi*, Discovery Pub, Delhi, 2005.

Parry M.L.: *Climatic Change, Agriculture and Settlements*, Dawson Folkestone UK, 1978.

Qystein V. Sjaastad: *Physiology of Domestic Animals*, International Book Distributing Co., Delhi, 2005.

Rao, S. Balaachandra: *Indian Mathematics and Astronomy*, Bangalore, Jnana Deep Publications, 1994.

Sharma, Rajni : *An Introduction to Plant Morphology*, Campus, Delhi, 2004.

Tripathi, Pramila : *Botanical Pesticides in the Management of Post-harvest Fruit Diseases*, Daya, Delhi, 2005.

Tyagi, I.D. : *Plant Breeding and Genetics at a Glance*, South Asian, Delhi, 2005.

Urton, Gary: *The Social Life of Numbers*, Austin, University of Texas Press, 1997.

Vanderplank, J.E.: *Plant Diseases: Epidemics and Control,* New York, NY, USA, Academic Press, 1963.

Vikas Chaudhary: *Entomology and Pest Management*, Navyug, Delhi, 2008.

Walia, S. and B.S. Parmar: *Pesticides, Crop Protection and Environment*, Oxford & IBH, Delhi, 1995.

Whealy K.: The Garden Seed Inventory, Decorah, Seed Saver Publications, 1988.

Yoshida T.: Cultivation of Citrus Genetic Resources for Evaluation of Characteristics, Tokyo, JICA, 1996.

Index

❑❑❑